Profitable Engineering

Transforming Technology Teams into Strategic Business Partners

Phil Clark

Published by: RYU Advisory & Media, LLC
ISBN: 979-8-9961169-0-4

Dedication

For my wife, Adrian —
the constant in my equations, the calm in my chaos, and the heart that makes it all worthwhile.
You've endured early-morning energy, late-night ideas, and everything in between as we've built a life filled with purpose, challenge, laughter, and love.
You remind me that the most meaningful work isn't written on pages or in code—it's lived, together, every day.

And to Bob Langan —
who recognized potential before I did and whose mentorship instilled the discipline to deliver and the courage to lead.
For that, I am forever grateful.

DEFINITIONS

Profitable Engineering (noun)

1. The discipline of leading technology not as overhead to be minimized, but as a strategic asset where efficiency drives *margin* and customer outcomes drive *growth.*
2. The synchronization of *Flow* and *Realization*: shipping fast with high quality, to deliver solutions that positively impact customer behavior and business results.

Table of Contents

Who This Book Is For

This book is written primarily for senior technology leaders responsible for the performance, direction, and business impact of modern technology organizations. That includes CTOs, CIOs, vice presidents of engineering, engineering directors, and other leaders accountable for turning technology investment into customer and business value.

It is also written for adjacent leaders and practitioners who help shape that system, including architects, product leaders, delivery leaders, and others working across digital transformation and modern product delivery.

Some of the concepts and terms used throughout the book are common within the technology industry, so certain ideas may appear before they are fully defined in later chapters.

This is a leadership book grounded in real-world experience. It is written for leaders trying to improve flow, strengthen alignment, make better investment decisions, and connect technology work to measurable outcomes.

Readers newer to these practices should still find the broader lessons accessible. Those already working inside technology organizations will likely recognize the language, tradeoffs, and organizational tensions immediately.

At its core, this book is about how leadership shapes the system through which technology becomes business value.

FOREWORD

Phil and I first met through the Value Stream Management Consortium, bonding over a shared focus on real-world challenges. Phil didn't just share his wins; he shared his struggles with misalignment, measurement traps, and the difficulty of applying systems thinking in volatile environments. One of these shared stories actually became a cornerstone case in my book, *Flow Engineering*—a classic example of the problems we try to address with Value Stream Management.

Phil had experience using the best tooling money could buy and still struggled to see systemic change. He spoke of metrics being gamified and weaponized. Of the "cost-center" framing placed on technology. Of the silos, conflicting incentives, and the dangerous disconnection between activity and outcomes.

Phil has seen the real world of IT organizations from both a consultant's perspective and from inside the executive suite. That is a rare perspective. As a result, each of our conversations was rooted in cold, hard truth.

One pattern that emerged in our conversations—which I have learned a great deal from—is Phil's practice of clearly stating three things:

1. "Here's where I struggle."

2. "Here's what I'm hearing—let me challenge you on that."
3. "Here's what I'm learning."

That is continuous improvement in real-time. It is the trait I look for in every leader I work with.

Phil and I often discuss a challenge facing every VP of Engineering: distinguishing signal from noise in a world drowning in frameworks. It's easy to feel like our systems are spiraling out of control.

Generative AI has made it possible to run very fast in the wrong direction. It used to be that ideas were cheap, but execution was hard. That is no longer true. Today, it is essentially free to generate plans, documents, code, and backlogs. If you want "average" perspectives, they are instantly available.

The Dunning-Kruger effect creates the conditions for us to grasp a nugget of AI-generated information and mistake it for a massive store of value. Phil and I have both shared stories of how we've mistakenly dug up nothing but dirt and run far in the wrong direction before arriving at our current understanding of how people actually work together.

One of the best protections from these dangers is what you're reading right now. First-hand wisdom distilled from decades of experimentation, unlearning, and relearning.

In 20 years of helping organizations optimize flow, I've observed one consistent pattern in transformation failures: Leaders narrow their focus, and then deploy that focus to the wrong target. Rather than looking across the organization, they zoom in on individuals. Rather than communicating a target *Outcome*, they hyper-fixate on *Output*.

We all know better, yet we do it anyway. It's too easy to look for our lost keys under the streetlamp. When I need to make sense of the chaos, I return to the core components found in this book: *Flow and Realization.* Where are we trying to get to, and how are things moving right now?

Using that framing, you can start to rebuild your systems of work.

As you read *Profitable Engineering*, I encourage you to adopt Phil's practice of continuous learning. Ask yourself: *Where am I struggling? What am I learning?*

The questions will help you step out of the moment to consider your own flow. The answers will guide you toward building a technology organization that is no longer a cost center, but a true strategic partner.

—Steve Pereira
Flow Advisor and Founder, Visible Flow Consulting
Co-author of Flow Engineering: From Value Stream Mapping to Effective Action

Like Phil, I've spent over three decades immersed in helping technology organizations improve their working practices. Our paths crossed at the Value Stream Management Consortium (VSMC) in 2022 (where I was founder and CEO/Chair) when Phil joined as a member and swiftly revealed himself to be a lifelong learner and champion experience sharer, speaking at our events and ultimately joining the board to direct the consortium's strategy. In the last four years, I have learned so much from Phil and now count him as part of the Flow Family at Flowtopia (the evolution of the VSMC)—our inner cohort of coaches, the most forward-thinking and advanced Flow Pros.

Phil is, in my experience, a rare breed. He's a leader who understands change needs to come from within. A leader who does not think he's learned everything but whose humility guides him to evolve as the world evolves around him. He is generous with his time and expertise, committed to his part in making the world a better place. This book of his, *Profitable Engineering*, is the perfect vehicle to present his extensive expertise on how to maximize value and accelerate flow in technology organizations.

I 100% align with the core of Phil's mission: that for the world to be the best place it can be, technology is a strategic human enabler, not an afterthought. The geek has inherited the earth. Software has eaten the world. But we are so many humans; there are so many systems; new technologies are continuously invented—the work we have chosen to do is not easy. It's complex and complicated. People resist change. External economic and geopolitical factors influence what we can achieve every day at our desks.

But Phil is patient and kind. He is curious, optimistic, and persistent. He's a deep thinker, hungry and respectful of others' battle-worn stories and euphoric successes. He has synergized what he's read and heard and tested it in real-world situations. He's invented and validated his own approaches alongside these, and now he's sharing what he knows.

Profitable Engineering provides a clear path, proven practices, and working philosophies that will help people navigate the challenges we face now and in the future; how to make the invisible visible. How to collaborate across organizational boundaries. How to mobilize people to experiment and embrace new ideas, methods, and technologies. How to measure what we do and why we do it. Phil understands how to connect flow insights to business risks and why outcomes and impacts are everything in a world where we are frequently distracted by labels and shiny new things. He has seen how realizing the value of the work that we do as technologists in the digital and

AI era is the key motivator, the driver of purpose, and the catalyst for change.

His patient determinism guides us through the steps to reshape thinking around a value stream oriented operating model, evolving roles to advise and strategize on flow optimization and value outcomes.

I invite you to turn these pages and let Phil's wisdom empower you; *Profitable Engineering* is more than a guide—it is an essential manifesto for anyone ready to lead their organization toward a more purposeful, flow-oriented, and profitable future.

—Helen Beal
Founder of Flowtopia

I have had the honor of knowing Phil for over 25 years. In that time, he has become my closest confidant and most trusted advisor. But this book is not born merely from our friendship; it is born from the trenches of our shared professional history, a history defined by solving the kinds of impossible problems that keep leaders awake at night.

Our journeys often ran parallel by design. Whenever one of us found a company with potential but lacking the critical engineering culture to unlock it, we made the call. Time and again, we recruited each other, confident in a shared philosophy that valued outcome over output and people over process.

We first collaborated during the volatility of the dot-com era at Evergreen Internet. Phil didn't just write code; he helped build the professional services organization that propelled the company forward. Long before terms like "continuous delivery" or "service-oriented architecture" were industry standards, Phil was laying their groundwork. When I became COO, he was the first person I turned to for counsel.

A defining moment in Phil's leadership, and one that convinced me of the necessity of this book, came the day after I promoted him to Director at Evergreen. We faced an abrupt financial crisis that required immediate, painful staff reductions. Phil insisted on notifying his people personally and standing with them through the process. It was a profound display of character and courage, proving he was not just a technology manager but a leader of people.

We later reunited at DHL and Parchment, tackling environments of immense complexity. At Parchment, I watched Phil blend Agile practices into rigid Waterfall structures and later evolve entire organizations using Flow Metrics and modern operating models. He didn't just implement frameworks; he engineered cultures where retention was remarkable and high performance was the default.

Today, Phil is a recognized expert who has successfully navigated the evolution of software engineering, from the early web to the age of AI. This book is a distillation of that journey. It is a rare look into the mind of a leader who understands that while technology changes, the principles of trust, flow, and value remain constant.

I hope you find this book as practical and transformative as I have found Phil's counsel over the years. It challenges old assumptions and offers a clear path for those ready to lead with both courage and care.

—Bob Langan

Former Senior Vice President of Engineering, Parchment Inc.
Former President, IT Services (Americas), DHL

Preface

Books accelerate knowledge; experience develops wisdom.

That idea has shaped much of my career and ultimately led to the insights shared in this book.

For years, colleagues, vendor partners, and fellow leaders encouraged me to put these lessons into writing. Eventually, I listened.

Experience is the engine of changed minds. As I immersed myself in books, conferences, conversations, and the day-to-day practice of new ways of working, the certainty I once carried gave way to a broader understanding of what modern software delivery actually demands. In the end, it wasn't expertise that transformed my leadership; it was openness.

This book isn't about what's right or wrong. It's about what many of us have learned by adopting modern practices, architectures, technologies, and cultures that help teams deliver value quickly and adapt just as fast. These approaches emerged through experience, experimentation, and evolution. Not dogma. They work today, but the next decade will ask new questions of all of us. The only enduring advantage is a willingness to stay curious and keep evolving.

For more than two decades, I have watched the software industry reinvent itself time and again. I lived through the race to digitize in the early Internet era, the shift to continuous delivery, and the rise of adaptive, team-centered models. Now, the industry stands at the threshold of the AI revolution.

But looking back, the defining feature of my career wasn't technology. It was unlearning and relearning.

Success at every stage required me to confront my own fixed mindset. The rigid hierarchies that served me well in the 2000s became the very things that held my teams back in the late 2010s. I discovered that the hardest part of transformation isn't adopting new frameworks. It is challenging deeply held beliefs about how work should be done.

None of this is easy. Today, you hear that Agile has failed, that Scrum Masters aren't necessary, or that small teams don't work. But those conclusions nearly always trace back to leadership, context, and implementation.

There is a world of difference between understanding the what and why behind these practices and simply forcing a new system onto an old mindset. Practices succeed when the architecture supports them, when leaders understand their purpose, and when teams are empowered to operate within the right constraints. Without that foundation, even the best frameworks collapse under the weight of misunderstanding.

I was not alone in this struggle. Time and again, I saw peers and organizations hit the same wall. I sat in rooms with senior leaders who were more rooted in legacy thinking than inspired by innovation.

Later, one of my direct reports relayed a colleague's side comment: "Phil probably finished another book." It wasn't a compliment or a complaint so much as a signal that when I brought in a new idea, I was trying to sharpen how we thought, decided, and executed. I was also met with the familiar refrain: "That's great in theory, but it won't work here."

Those moments were both frustrating and clarifying. They taught me the barrier to progress wasn't a lack of tools; it was mindset and a lack of shared understanding.

That realization became the foundation of why I wrote this book: to help leaders rethink engineering effectiveness and impact and make software delivery a strategic business advantage by shifting technology from a cost to control into a strategic partner tied directly to business outcomes.

I lead with two triplets: Outcomes, Flow, Visibility. Accountability, Quality, Systems. They're the guardrails I come back to when pressure is high, so we don't revert to activity, heroics, or certainty theater.

I have seen firsthand how linking technology investments to measurable results transforms not just delivery, but culture. When leaders connect purpose to performance, and teams understand the "why" behind their work, engagement rises, and friction fades. And as I often say: fast flow happens when friction fades.

This book comes from experience earned inside the work. It reflects decades of lessons shaped by the organizations I've served, the mistakes I've made, and the conversations I've had with peers equally committed to rethinking the intersection of technology and business. I've learned as much from missteps as from successes.

Through advisory board work, executive roundtables, and workshops, I've learned that the most significant breakthroughs rarely come from frameworks or certifications. They come from honest conversations, in trusted spaces, where leaders let down their guard, share what isn't working, and challenge each other to think differently.

The giants of our industry have also shaped my perspective, from the thinkers behind the IT Revolution to the authors of *The Goal*, *The Phoenix Project*, *Project to Product*, *Accelerate*, *The DevOps Handbook*, *Flow Engineering*, *Team Topologies*, and so many others. This book stands on their shoulders while aiming to illuminate the next step in the evolution: the practical, profitable application of these principles in the real world.

Transformation is not a destination; it is a discipline, a continuous loop of learning, unlearning, and relearning. Over the last seven years, I've returned to these ideas to stress-test and refine the cultures, practices, and operating models detailed in these pages.

This is both a reflection and a roadmap. My aim is to show how clear priorities, shared purpose, and strong leadership can transform technology from overhead to advantage.

If you've ever felt the tension between what your organization says it values and what it actually rewards, you're not alone. I wrote this book for leaders under pressure, short on time, and trying to make sense of why talented teams still struggle to deliver with clarity. Often, the problem is not talent. It is that the system turns decisions into drag, priorities into noise, and delivery into a grind.

If this book helps even one leader see technology differently and inspires them to lead with purpose, courage, and curiosity, it will have fulfilled its purpose.

ACKNOWLEDGMENTS

This book reflects years of experience, experimentation, learning, and trusted collaboration that shaped my path.

Holly and Bobby—your thoughtful feedback, patience, and steady guidance made the editing process both productive and meaningful. Thank you for helping bring this book to its best.

To the teams at Parchment, past and present—you are the heartbeat of this story. Every deployment, every retrospective, every win, and every hard lesson shaped what's here. You turned frameworks into reality and ideas into impact.

Steve Pereira—thank you for your generosity in sharing the practice of Flow Engineering across organizations of every size and stage. Our conversations and collaborations challenged me to see flow not just as a set of metrics, but as a leadership mindset.

To my friends, collaborators, and peers across the Agile, DevOps, and Value Stream communities—thank you for challenging me, sharpening my thinking, and raising the bar.

Special appreciation to Helen Beal and the Value Stream Management Consortium community whose work now lives on through Flowtopia.io, for keeping the dialogue alive and the ideas evolving.

To the authors and thought leaders who shaped my perspective—thank you. This book draws on foundations such as *The Goal* (1984), *Turn the Ship Around!* (2012), *Mindset* (2006), *Think Again* (2021), *The Coaching Habit* (2016), *The Lean Startup* (2011), *Continuous Delivery* (2010), *The Phoenix Project* (2013), *Accelerate* (2018), *The DevOps Handbook* (2016), *Project to Product* (2018), *Team Topologies* (2019), *Making Work Visible* (2017), *Sooner Safer Happier* (2020), *Escaping the Build Trap* (2018), and *Flow Engineering* (2024), among many others..

To the broader community of practitioners—engineers, product leaders, delivery professionals, and people who share what they learn—thank you. Progress continues because you experiment, challenge norms, and make the work visible.

And to the readers and leaders who picked up this book—thank you for your curiosity. If something here helps you see your teams, your systems, or yourself a little differently, then this effort was worth it.

Let's keep learning from one another. Let's keep improving how technology serves people and how leadership serves purpose.

Our shared journey is far from over—but it's already underway.

INTRODUCTION

All work shows up as a cost on the ledger. The difference is whether that cost creates leverage.

Execution is intelligence. It's the operating rhythm where decisions turn into outcomes.

This book is about building that rhythm—making invisible knowledge work visible, turning the delivery system into a living dashboard, and engineering a system of visible priorities, feedback, and accountability that reduces friction so value can flow measurably, repeatedly, and without heroics.

Nobody wakes up hoping to spend the day chasing approvals, sitting in status meetings, or shipping work that doesn't matter. People want to contribute. They want to build, solve, and improve. But when purpose is unclear and systems are rigid, even great teams spend more time navigating bureaucracy than delivering value. Over time, that friction doesn't just slow delivery. It drains energy, erodes trust, and trains the organization to expect less.

And in today's world of Agile, DevOps, automation, continuous delivery, and now AI, speed and adaptability aren't differentiators. They're the baseline. Competitors are already shipping in smaller batches, tightening feedback loops, learning faster, and compounding advantage. If your organization can't match that cadence, you're not just moving slower. You're falling behind.

The fundamental constraint isn't capability. It's friction.

Many organizations already know the practices. They've heard the frameworks. They've adopted some of the tools. What holds them back isn't a lack of intelligence or effort. It's the weight of bureaucracy, rigid hierarchies, and legacy leadership reflexes that create drag across the system. Friction shows up as:

- decisions that take weeks instead of hours
- priorities that change without explanation
- work that stays invisible until it's too late to steer
- teams measured by activity instead of outcomes
- alignment that depends on heroics, not visible priorities and ownership

This is why the same practices can look transformative in one company and feel like theater in another. It's not the framework. It's the system around it.

Fast flow happens when friction fades.

You may never remove all friction, but when you measure it, name it, and deliberately reduce it, progress can feel effortless. Work moves. Teams regain ownership. Trust rises. Outcomes become consistent.

From Cost Center to Strategic Partner

Nowhere is this disconnect more visible than in how organizations view Technology.

Many still treat Technology as a cost center: a function that executes predefined work, maintains platforms, and minimizes expense. That mindset produces a reactive culture: compliance over curiosity, efficiency over effectiveness, output over outcomes.

But in a digital economy, that view becomes a liability.

Technology is no longer just a support function. It is the backbone of strategy, growth, and customer value. The organizations that thrive treat Technology as a strategic partner, not because they romanticize engineering, but because they recognize the economics:

- Profit sustains the mission.
- Efficiency fuels innovation.
- Cost discipline creates freedom.

From Cost to Profit When Technology Is the Business

Technology stops being "just a cost" the moment it proves its worth in business terms.

Every platform investment, architecture decision, and delivery improvement must tie back to outcomes the business can recognize: profit, efficiency, risk reduction, or cost-to-serve improvements. And when Technology is the product, as it is in a SaaS company, the stakes rise even higher. Here, technology doesn't just enable the business. It is the business.

Profit depends on delivering solutions customers will pay for and stay with.

Efficiency depends on shortening the path from idea to customer value.

Cost discipline depends on eliminating waste, debt, and friction across the flow of work.

When those forces align, technology stops being overhead. It becomes an advantage because no matter how noble the mission, no company can change the world unless it first sustains itself.

Profit funds purpose.

What This Book Will Do for You

This book is written with one outcome in mind: to give you a clearer way to see the system, the models to improve it, and the leadership discipline required to turn Technology from a cost to manage into a strategic engine tied to measurable business results.

Specifically, this book will help you:

- Shift the logic: Move from project-cost thinking to outcome-and-economics thinking.
- See the system: Understand how architecture, team design, and operating model fit together so improvements compound rather than compete.
- Make invisible work visible: Turn knowledge work into signals, and the delivery system into a living dashboard you can actually steer.
- Measure what matters: Replace vanity metrics with steering metrics that connect flow to customer outcomes the business respects.
- Trade certainty for confidence: Stop demanding "perfect estimates" that force teams to lie, and start managing software as a series of probabilistic bets.

- Lead the transformation: Recognize legacy reflexes in the mirror and replace them with clear priorities, constraints, and empowerment.

THE CONVERGENCE MODEL: FLOW, FUNDING, REALIZATION, AND AI

Profitable Engineering is the synchronization of Flow and Realization.

Flow is how work moves from idea to customer value. Realization is the measurable business return from that work. In modern organizations, those two only stay connected when leaders intentionally design the four forces to work together:

- Flow is the operating engine. It makes work visible, exposes waiting and friction, and turns "we're busy" into evidence about where time is being lost.
- Funding is an incentive design. It determines whether the organization funds products for long-term learning or continually resets momentum with project-era governance.
- Realization is the destination. We stop celebrating "features shipped" and start measuring "outcomes achieved" verifying that our investments actually changed customer behavior or business results.
- AI is the capacity multiplier. It accelerates whatever system it is placed in, which means it amplifies shared understanding when the system is healthy and confusion when it is not.

This book builds a single executive operating model from these threads:

- Flow is how you deliver.
- Funding is how you align incentives.
- Realization is how you measure value.
- AI is how you scale capability.

That convergence is how Technology stops being managed as a cost center and starts being led as a strategic partner.

How the Journey is Structured

Part 1 surfaces the legacy mindset that keeps Technology trapped as overhead. It replaces metric theater with steering signals shifting from estimates to bets, confidence, and evidence that connects flow to outcomes.

Part 2 focuses on the practical transformation: how operating models, team design, architecture, and flow create repeatable delivery.

This book is not a comprehensive implementation manual for every practice supporting modern digital product delivery. That is intentional. Its purpose is to help leaders see the larger system more clearly and understand how flow, outcomes, operating models, team design, and strategic alignment connect.

For readers who want to go deeper, the appendices serve two different purposes. Appendix C is the "how we approached it" appendix. It shows how we identified products and value streams using domain thinking, inputs and outputs, and other signals inside the system.

Appendix D, *Suggested Reading and Companion Practices for Modern Digital Product Delivery*, is the "go deeper here" appendix. It points to companion practices and suggested reading for adjacent disciplines such as User Needs Mapping, Wardley Mapping, Domain-Driven Design, architecture visualization, Team Topologies, and Value Stream Management.

Part 3 connects it to the outcomes leaders care about: revenue, retention, and profitability. It introduces the financial spine that governs outcomes over time and unites Flow, Funding, Realization, and AI into a single operating model.

It treats AI as a capacity multiplier anchored in literacy, guardrails, and measurement.

It also addresses Talent Divergence, where AI fluency can mask weak fundamentals, and what to look for in modern engineering talent.

Modern delivery will not be unlocked by adding more ceremonies. It requires different leadership and better system design.

When strategy shifts from a static slide deck to a living dashboard, momentum accelerates, trust strengthens, and outcomes no longer hinge on heroics. That is what rethinking success in engineering leadership looks like: not merely measuring how fast you ship, but how clearly you deliver impact.

Let's begin.

Key Terms Used in This Book

Before going further, it is worth defining two terms that appear throughout this book. Both are widely used, but they are not always used consistently. There are several important ideas woven through this book, but these two show up so often and carry so much of the argument that they are worth defining up front.

Value Stream Management is the practice of making work visible from idea to customer value so leaders and teams can see where time, effort, and friction are being lost. In this book, it is a leadership discipline for improving flow, exposing constraints, and strengthening the connection between technology work and business outcomes.

The Product Operating Model organizes teams, funding, and decision-making around long-lived products, customer value, and continuous learning rather than focusing solely on temporary projects. In this book, it matters because it creates the conditions for teams to stay close to outcomes, adapt as they learn, and build momentum over time rather than repeatedly reset around short-term delivery cycles.

Many excellent resources go much deeper into the practices and implementation details behind these disciplines. This book focuses on why they matter in leadership, how they shape the environment around teams, and how they help connect technology investments to business results.

PART 1: THE LEGACY MINDSET

If you are not unlearning, relearning, and rethinking your understanding, you are not just behind; you are the bottleneck.

In this section, we dismantle the financial and leadership reflexes that trap organizations in the past. We examine why the "Cost Center" model fails in a digital world, why "efficiency" is often a trap, and how the habits that made leaders successful yesterday are the very things holding their teams back today.

1
The Cost Center Paradox

"There is nothing so useless as doing efficiently that which should not be done at all." — Peter Drucker[1]

The Productivity Trap

Liz Wiseman, in her book *Multipliers*, describes two leadership patterns: Multipliers, who draw out and expand the intelligence around them, and Diminishers, who unintentionally suppress capability and leave it on the table.[2]

In modern product and engineering environments where complexity outpaces any single person's ability to "know and tell", organizations can't afford leadership models that slow decisions, narrow ownership, and reduce psychological safety. When leaders create high-trust, low-blame environments, performance improves; when fear and blame dominate, performance suffers.

This is why many "Agile failed here" or "DevOps isn't working" stories are misdiagnosed. The practices are blamed, but the real constraint is the operating system of leadership. You can rename meetings, adopt a framework, and install new tooling. Yet still, run a Diminisher environment where decisions bottleneck at the top, metrics become surveillance, and teams optimize for approval instead of outcomes.

In those conditions, Agile and DevOps don't fail. They never get a chance to work.

Why it matters: Methods don't create performance. Leadership and culture determine whether modern practices become a force multiplier or simply a new vocabulary for old dysfunction.[3,4]

One of the most evident signs that legacy thinking persists is how senior leaders want to measure technology performance. At Flowtopia 2023, an industry conference focused on flow and value stream management, I shared a story that crystallized this point.

After years of transformation—Agile adoption, DevOps practices, cross-functional teams, and early Value Stream Management efforts to make the flow of work visible from idea to customer value—the same request kept surfacing, in my organization and in others I've since advised: "Show us productivity units."

The reasoning was understandable. If sales and support performance could be measured in simple units, why couldn't engineering? More individual productivity "units" should equal more value. This approach is a textbook legacy reflex. It reduces knowledge work to transactions and ignores the complexity of software delivery. Worse, it sends the wrong signal to teams. When engineers know they are judged by individual output, they optimize for activity rather than outcomes: more lines of code rather than solving customer problems, more tickets closed rather than reducing cycle time.

Research echoes this risk. Empirical studies of software delivery performance and engineering productivity show that activity-based measures such as commits, story points, and ticket counts primarily reflect effort rather than customer value and can distort behavior when used as performance targets.[5-7]

The pressure for individual productivity metrics is more than a misunderstanding. It is a leadership reflex inherited from an era when predictability mattered more than adaptability. I have encountered this reflex at multiple points in my career, and peers across the industry report the same pattern.

That reflex didn't disappear. It simply changed shape.

Technical debt isn't only messy code. Another form of debt accumulates more quietly: the cost of operational blindness. When leaders can't see where work waits, why decisions stall, or whether releases change customer behavior, they end up managing by stories instead of signals. That blindness becomes expensive in much the same way code debt does: it creates rework, delays learning, and forces late-stage tradeoffs when the only remaining lever is pressure. The quickest way to reduce that debt is not another status meeting; it is making the system visible enough to govern.

When leaders feel exposed, they reach for certainty: dates, commitments, and "productivity" proxies that make software feel controllable. But software change is rarely repeatable work. It's learning work, and the plan gets rewritten as reality shows up.

If you want predictable outcomes, you don't start by demanding better estimates. You begin by defining what matters, why it matters, and what you're willing to stop. In Chapter 4, we'll revisit estimation and predictability directly replacing promise-based planning with confidence-based forecasting grounded in flow.

To move forward, I had to reframe the discussion. Instead of defending Agile or DevOps, I introduced Flow Metrics, which gave us visibility into how work moved through the system, where it slowed, and how delivery performance changed over time. These metrics focus on team performance and system efficiency rather than individual output. Flow time revealed that delays came not from engineering inefficiency but from business decision bottlenecks. Flow load showed that overcommitment, not a lack of effort, was the real cause of missed timelines.

The turning point came when I tied these insights to outcomes executives already cared about: business risk, customer satisfaction, and growth potential. In practice, overloaded teams were not just an engineering issue; they increased the risk of delays, quality problems, and missed revenue opportunities. Long approval cycles weren't just frustrating; they represented missed opportunities in the market.

The lesson is clear: leaders who shift from outdated metrics to outcome-driven measures will find new ways to unlock value and inspire their teams to focus on what truly matters.

The Origins of Control

This view of technology as a cost to be managed is not just an organizational habit. It has deep roots in the way business has been measured for more than a century. The management systems built to control cost and predict work also shaped how leaders expected technology teams to behave.

I lived through the era when many of those assumptions hardened into practice.

It is well past midnight in the Chandler, Arizona office. The only light comes from the glow of CRT monitors. My empty soft drink cup sits beside me as lines of code and log files scroll across the screen. We had been building this site for months, waiting for this exact moment. When the first online order finally appears, a cheer erupts across the room. High-fives fly. We have just brought another legacy catalog business into the new frontier of online commerce.

These are my memories from Evergreen Internet, my first real job as a software engineer during the late 1990s e-commerce boom. The team helped companies at the time, like The Sharper Image, The Swiss Colony, and Midnight Velvet, put their catalogs online before "e-commerce" was a household word. It felt like we were part of something bigger than ourselves, bringing decades of print catalogs into the digital world.

We built with the best tools the era could offer. Java was still finding its footing, XML fueled our data broker, and we were wiring our ECential™ commerce platform into back-end AS/400 systems so an online order landed in the same queue as a phone order.

Nights and weekends often blurred together. Not because we had to be there, but because we wanted to be. Every launch felt like a victory for the web itself.

But Evergreen was also my first real exposure to the waterfall software development life cycle. Business analysts and designers would spend months creating exhaustive requirement documents. Engineers like me would then produce equally detailed technical design documents before a single line of code was written.

The process was rigidly linear: requirements, design, build, test, release. Quality assurance cycles stretched for weeks or months. And when launch day finally arrived, we crossed our fingers, hoping the thing we had built six months earlier still matched what the customer needed.

Back then, this felt like success. Discipline, process, and documentation were the markers of professionalism. The more detailed our requirements and design docs, the safer we felt. What I did not realize then was how much risk we were building into the system.

By the time we shipped, customer needs had often shifted. We were not validating the value along the way. We were betting everything on a single release moment.

There were moments when those risks became obvious. I remember how often, weeks or even months into a build, the ground shifted beneath us. Project managers, working closely with customers, would return with changes weeks or months into the effort, sometimes because the catalog system had been updated, other times because the customer had a new idea. What we had already documented, scoped, and sometimes written had to be undone and rewritten. Scope changes weren't the exception; they were common.

Years later, the model itself guaranteed this kind of misalignment. We were not learning along the way. We were waiting, hoping, and betting.

Evergreen's story was both exhilarating and humbling. The dot-com boom filled our halls with optimism, but the crash that followed forced us to adapt.

By the time I left in 2002, Evergreen's pioneering era was drawing to a close. Still, the habits it instilled—long planning cycles, heavy documentation, and a belief in control—shaped the way I thought about software delivery. Those instincts would carry forward into my next chapter at DHL, where discipline and control were elevated into a full-blown operating model.

DISCIPLINE AT SCALE: THE DHL YEARS

Joining DHL in Scottsdale in 2003 felt like stepping into the big leagues. The scale was staggering. This was one of three global data centers powering DHL's worldwide shipping network in continuous, follow-the-sun cycles. At that time, Deutsche Post DHL employed more than 320,000 people globally, with 70,000 working under the DHL brand alone.[8]

My early days were shaped by enterprise practices common during that period: long lead times, exhaustive requirements documents, tightly choreographed release processes, and planning sessions that often stretched for days.

Quality assurance meant sitting in someone else's queue, waiting for approval before anything moved forward.

My first assignment dropped me straight into that model, taking over an ocean logistics system (a Java EJB monolith on an application server) acquired through the recent Danzas Ocean merger, which had to be resilient and performant at an enterprise scale.

Later, at DHL, those same practices were scaled and formalized across a massive global enterprise. What had been habits at Evergreen—detailed documentation, rigid process, control over every step—became codified into frameworks and certifications that defined how work was scoped, tracked, and measured.

The company itself was in an enterprise hyper-growth phase, fresh off acquisitions like Airborne in the U.S. and Danzas in global freight. Integration ran on a strict playbook, with a ninety-day timeline to integrate systems, and 'Day 100' posters lining the walls marking the deadline for full integration.[9] Across Express, Ground, and Ocean, the practices were the same: coordination across dozens of teams, governance boards controlling every change, and long timelines before a release reached production.

Even the physical space mirrored the model, with tables of developers crowded into an open floor, glass-lined offices circling the perimeter, and a first-floor Network Operations Center (NOC) featuring giant screens that tracked network health and global news updates. This experience was a waterfall software development lifecycle in motion at scale, the way enterprises of the time managed risk, quality, and predictability.

Frameworks like PMI, Prince2, ITIL, and Six Sigma were not just methods. They were the culture. Every project had a charter, every service change a defined request, and every improvement initiative a set of measurable targets. PMI emphasized detailed project plans and work breakdown structures. ITIL governed service changes and incident response with an insistence on predictability. Six Sigma drove efficiency by pushing us to identify variation and eliminate waste with precision.

In practice, this meant work was rarely ambiguous. We had forms, approvals, and processes for every stage. A service change might pass through multiple committees. A project update meant reviewing earned value reports, schedule variances, and utilization charts. It was discipline taken to an industrial scale.

I recall one change request in particular. It was just a minor update, each requiring its documentation and sign-off. By the time it was approved, weeks had passed. The delay was not about complexity. It was about proving compliance. That was the culture: every signature, every approval, every earned-value chart became evidence that IT was managed, not that it was creating value.

Meetings often reflected the same mindset. I recall sitting with senior managers as we debated whether a project was 63 percent complete or 70 percent complete, while the actual customer impact of the work barely registered in the discussion. Leaders were rewarded for their ability to manage variance and enforce reporting discipline. What mattered was that you could prove the system was under control.

This pressure to control variance created what we now call the "Watermelon Effect."

In the monthly reviews, every project was reported as Green (on track). Green meant you were a good manager; Green meant you were in control. But because we were measuring activity rather than outcome, the Green status was often a shell. Inside, the project was Red—riddled with unproven assumptions, integration risks, and technical debt that wouldn't be discovered until the testing phase, months later.

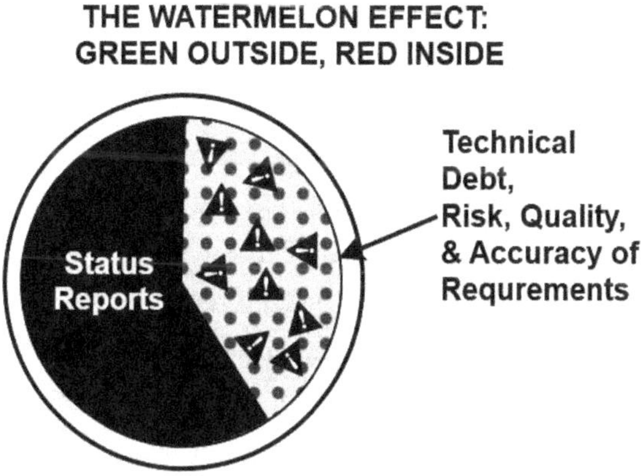

Figure 1.1 Author Illustration: *The Watermelon Effect*

Scope changes were inevitable once users finally saw the software, and those changes wrecked the rigid plans we had spent months perfecting. I watched thousands of dollars evaporate in rework during the final weeks of these "Green" projects. When reality finally broke through the reporting theater, we were forced into crunch mode to save the timeline. The cost center model didn't mitigate risk; it obscured it until the very end.

The incentives reinforced this worldview. Utilization reports, earned-value dashboards, and red-amber-green status charts became the scoreboard. A manager might be praised for hitting efficiency targets even if their teams delivered little new value.

In one review cycle, I experienced my rating tied to variance-control and resource-allocation metrics rather than business outcomes. In that executive frame, success meant predictability, compliance, and efficiency against plan rather than whether our work created meaningful business results.

In fairness, this mindset had real benefits for a global logistics enterprise that moved millions of packages each day. Strict change controls and precision-driven frameworks kept critical systems stable. Packages did not get lost because someone skipped a step. Customers received reliable service because every change was documented, tested, and approved. From a business continuity standpoint, it worked.

But those same benefits came with hidden costs. Most changes had long tails; developers were not seen as creative professionals building the future of digital logistics. They were 'resources' to be allocated and optimized. Projects were measured by compliance and efficiency rather than impact. The cost-center mindset was not only present; it was institutionalized.

The Industrial Hangover

Why does this mindset persist? The very idea of "cost centers" originated in industrial-era cost accounting.

At the turn of the 20th century, manufacturers needed to track wages, overhead, and production costs with precision to keep factory operations predictable. The model made sense for physical goods. Managers optimized for efficiency, reduced variation on the assembly line, and defined success by controlling costs rather than creating new value.

When IT entered the picture decades later, it was slotted into those same accounting categories. Budgets were allocated as fixed expenses, reviewed annually, and judged by whether they stayed within budget. Technology was not seen as a growth driver but as a service function, something you funded to keep the lights on, much like facilities or payroll.

The CapEx vs. OpEx Trap

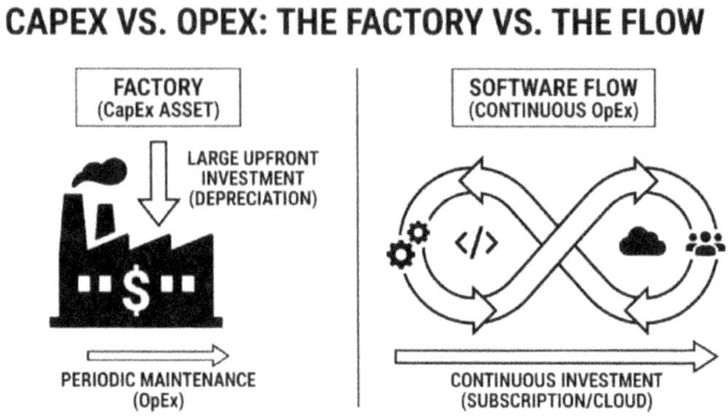

Figure 1.2. CapEx vs. OpEx: The Factory vs. the Flow. AI-assisted illustration.

This accounting history explains a tension that still frustrates CTOs today: the battle between Capital Expenditure (CapEx) and Operating Expenditure (OpEx).

In the industrial model, building a factory is a Capital Expenditure, an asset that depreciates over time. Running the factory is an Operating Expense. CapEx is often favored in financial planning because accounting treatment allows it to improve EBITDA (Earnings Before Interest, Taxes, Depreciation, and Amortization).

Consequently, the Cost Center mindset forces software into the "Factory Building" box. We are forced to define large, capitalized "projects" with start and end dates so Finance can depreciate the asset. Continuous delivery, where we perpetually improve the software, looks like OpEx (overhead) to a traditional CFO.

This is why your Finance partner resists funding long-lived teams. They aren't trying to be difficult; they are trying to follow accounting rules written for steam engines. Leaders must understand this: we aren't just fighting culture; we are fighting the Generally Accepted Accounting Principles (GAAP).

As Jon Smart notes in *Sooner Safer Happier*, this legacy is why so many organizations still categorize IT as a cost center today, even though digital capabilities have become central to competitive advantage.[10]

Melissa Perri, in *Escaping the Build Trap*, describes how this mindset warps incentives. Leaders focus on outputs (features shipped, hours worked, capacity consumed) rather than outcomes like revenue growth, customer retention, or time-to-value.[11] The result is an illusion of productivity that looks efficient in the short term but rarely delivers sustained business impact.

I'd seen this pattern firsthand. At DHL, success was measured in utilization rates, adherence to scope, and variance from plan. Meetings often revolved around whether the budgets and timelines assigned to the work aligned, not whether the work created value for customers or the business. Leadership reviews celebrated managers whose teams were entirely "billable," even when the work produced little customer impact.

When leaders are rewarded for predictability, they naturally double down on control. When the scoreboard is utilization, they focus on maximizing hours. When variance is punished, adaptability looks like failure. Over time, the mindset becomes ingrained not just in individuals but in the culture itself.

This logic might have worked in an era of stable, repeatable production. But in the digital era, where each new initiative has unknowns and requires learning along the way, it creates fragility. Organizations that cling to this mindset risk being efficient at the wrong things: shipping outputs that do not matter, meeting budgets while missing opportunities, keeping systems stable while competitors pull ahead.

The Unlearning: The High-Growth Paradox

Looking back, I often describe my journey in stages. Evergreen was my bachelor's degree—my introduction to building software. DHL was my master's degree—teaching me process, governance, and scale. But Parchment was my PhD—a crash course in high-growth leadership that forced me to question everything I thought I knew.

The stakes at Parchment were entirely different. We faced a unique market constraint: the product served the academic industry, which resisted disruption during the school year. This meant we effectively only had a three-month window for major releases.

That academic calendar created a dangerous comfort zone. Early in my tenure, we shipped massive batches twice a year. Between those events, code sat in long-lived feature branches, accumulating risk and conflicts.

To break this cycle, we pushed hard into modern engineering practices. We moved from "Water-Scrum-Fall"—where we treated sprints like mini-waterfalls—to true Trunk-Based Development and Continuous Delivery.

We re-architected our platform, breaking a massive legacy monolith into a distributed microservices system. This architectural decoupling allowed us to dismantle our rigid functional silos and reorganize into small, cross-functional, self-managed teams.

Technically and structurally, we were transforming. We were moving faster than ever. But while the software delivery model and the architecture had changed, my mental model had not. I was still trying to manage this high-speed, distributed system with the static levers of the past.

If one team had more work than it could handle and another seemed to have capacity, my reflex was to move engineers around. To me, it felt efficient. In my head, I was "optimizing resources."

I was ignoring the critical aspects of team cohesion and development, treating high-performing units as mere collections of interchangeable parts. I was unknowingly mixing two operating models: the functional, capacity-driven structure of my past and the cross-functional, outcome-driven model we were trying to build.

As I would soon learn, this reflex to control "resources" would become one of the more complex barriers to our transformation. Eventually, I had to learn to prioritize team stability over short-term utilization.

The Illusion of Progress

Yet, even with those personal victories, the organizational pull toward the old ways remained strong.

Even after we stabilized teams and started delivering to production daily, I realized we were still operating in a feature factory model—shipping features and meeting targets without aligning the work to clear outcomes. We focused on output, meeting targets, shipping features, and responding to requests, without ever discussing anticipated results. We celebrated velocity instead of value.

Eventually, I had to learn to prioritize team stability over short-term utilization. I stopped moving people to the work and started bringing the work to the teams.

This is the Cost Center Paradox: although the tools had modernized through Agile and DevOps, and team stability had improved, the financial and strategic mindset often had not.

Leaders were still asking for "productivity units" in a world that demanded outcomes.

Reality Check: The CFO's View

You cannot simply declare, "We are no longer a cost center." To a CFO, until you can prove a direct link between a release and revenue (or retention), you are an expense. The transition from Cost Center to Strategic Partner is not a declaration; it is a journey of proving value. We will cover the metrics that demonstrate this in Chapter 4.

In many organizations, the shift from 'cost center' thinking to a true strategic-partner model is incremental. Even when delivery metrics improve, funding and governance habits can lag. This combined with financial pressure, legacy behaviors often reappear.

KEY TAKEAWAYS

- *Mindset defines success.* Treating technology as a cost center traps organizations in efficiency games, while outcome-driven leadership unlocks value and growth.
- *The Trap of Efficiency.* Being "disciplined" at scale (like DHL) creates stability but can suffocate value if the goal is compliance rather than customer impact.
- *The Measurement Error.* Leaders who ask for "productivity units" force teams to optimize for activity (output) rather than business results (outcomes).
- *History repeats.* The habits of the industrial era (cost accounting) have been copied and pasted into the digital era, creating fragility. Advantage no longer comes from controlling expenses; it comes from adapting quickly and delivering value continuously.

Recognizing the Cost Center trap is only the first step. The harder challenge is acknowledging that leaders often keep it alive through habits inherited from an earlier era, when command-and-control management was considered normal. I believed I had left that mindset behind when I adopted Agile.

I was wrong.

As we will see in the next chapter, the most significant barrier to transformation wasn't the finance department or the legacy code—it was the legacy thinking I was seeing in the mirror.

2
THE LEGACY MIRROR

As Alvin Toffler observed, the defining skill of the modern era is the ability to learn, unlearn, and relearn.[1]

The Hidden Barrier

Even after we began adopting modern practices in that period of transformation, I found the most challenging part of the transformation wasn't learning new methods. It was unlearning the habits and reflexes that once defined success. That is where legacy thinking takes hold.

Transformation often feels like learning a new language. You pick up the vocabulary of Agile or DevOps, but your accent gives you away. By 2018, the teams had made real progress: breaking down a monolith into microservices, reorganizing from functional silos into cross-functional teams, and shifting from infrequent releases to a weekly cadence.

On the surface, it appeared that we were transforming. I was fluent in the ceremonies of modern delivery, yet the way I spoke, acted, and measured still carried traces of my old management mindset. Legacy thinking has a way of sneaking back in, even when you believe you've left it behind.

And it wasn't just me. Other groups were still learning what agility meant in practice. Product teams struggled to balance vision with capacity and the pressures of iteration. Senior stakeholders wrestled with what transformation meant for customers who had once demanded absolute stability: no changes during the school year, no disruptions to established processes.

These tensions revealed that legacy thinking was not confined to individuals. It lived in the organization's structures, its culture, and its relationships with customers. Which raises the deeper question: if transformation is underway, why do these reflexes keep resurfacing?

To answer that, we need to understand why legacy thinking persists.

Why Legacy Thinking Endures

If these outdated patterns were simply ignorance, they would be easy to fix. Train leaders in agile, DevOps, or Value Stream Management, and the problem would disappear. But it isn't ignorance that makes these patterns so stubborn.

In my experience during that period, legacy thinking persisted because it is embedded in three places: the successes that once shaped leaders' careers, the systems that keep reinforcing those habits, and the reflexes that resurface when pressure mounts.

For decades, command-and-control approaches delivered results in manufacturing, where repetitive assembly-line work made effort and output predictable.

When software delivery became an established corporate function in many organizations during that era, leaders carried those same command-and-control assumptions into it. Work was planned as large, sequential projects. Functions sat in silos, handoffs landed in queues, and delays accumulated at every stage. Productivity was measured by individual output, and success was judged by hitting scope, time, and budget, even though those estimates rarely held.

Agile arose to address the limits of this model, but the reflexes of the past remain. Software is different. Every change introduces risk and unknowns, and the degree of those "unknown unknowns" makes precise prediction elusive. Command-and-control didn't create genuine predictability. It created the appearance of control.

Yet careers were built on meeting plans and budgets. Those expectations hardened into leadership reflexes that many executives still carry forward today, even in organizations that claim to be modern.

These reflexes don't vanish when new technologies appear; they find new outlets.

AI provided a timely example. In my early conversations in 2025, many executives approached AI the same way they had once approached outsourcing or automation: as a cost-cutting lever. The instinct was to trim headcount or squeeze more output from the same teams rather than to ask how AI could improve workflows, enhance the developer experience, or accelerate delivery.

Framed this way, AI adoption becomes another expression of cost-center logic. It becomes a modern tool used to repeat an old pattern. Modern leaders must unlearn this reflex and reframe AI as a *system enabler*, one that amplifies value when paired with healthy practices, rather than a substitute for them.

The Personal Struggle: Resources vs. People

I had to confront this truth personally as legacy reflexes showed up in my leadership. Some were obvious. Others were so ingrained that I didn't notice until someone called me out. Unlearning was not a single moment of insight, but a series of uncomfortable realizations, each forcing me to let go of practices that had once felt responsible.

One of the hardest habits to break was my instinct to move people between teams, or even shift whole teams to different products whenever a project needed more capacity. Early in my career, that felt like good management: rebalance headcount, protect the deadline. On paper, it looked efficient. In practice, it disrupted trust, eroded cohesion, and created new dependencies that slowed delivery. What looked like optimization was really a reflex: treating people, or even entire teams, as interchangeable.

Every move came at a cost. Relationships built over months were disrupted. Product knowledge was diluted. Teams left gaps behind, and the people who remained had to relearn the context that had quietly walked out the door. My intention was acceleration, but the system-level result was often the opposite.

If one team had more work than it could handle and another seemed to have capacity, my reflex was to move engineers around. To me, it felt rational. I treated work like a puzzle, moving the pieces until the picture fit. In my head, I was "optimizing resources."

That language is what exposed me.

A Scrum Master started calling me out every time I used the word "resources." The first time, I brushed it off. She kept reminding me, "They're people, not resources." At first, it sounded like semantics. Then it became a mirror.

One particular team review is still vivid. We were mid-sprint, and I was outlining why two engineers needed to be reassigned to help another project hit a deadline. I laid out the reasoning: one team had too much work in flight, and another seemed underutilized. It made sense on the spreadsheet.

The Scrum Master cautioned that reassigning people mid-sprint would break flow. It would force context resets, rebuild rhythm from scratch, and slow everyone down. For the first time, I realized they were not just hearing my logic. They were living the consequences of my decisions and the friction those decisions created in the effort to build cross-functional, self-managed, long-lived teams that could own delivery end-to-end.

My language wasn't just shaping perception. It was reinforcing a worldview I thought I had outgrown. The more I reflected, the more I realized language isn't incidental. It's a window into mindset. And if my words betrayed a belief that people were interchangeable, my actions inevitably followed.

What began as an awkward correction became one of the most important leadership lessons of my career: words matter because they reveal what you believe. Changing my language was the first step toward changing how I led.

It took me years to fully grasp that the hidden cost of these moves was far greater than the short-term benefit. High-performing teams aren't simply collections of individuals. They are long-lived groups that build trust, rhythm, and shared understanding. By treating people as pieces to be rearranged, I was reinforcing the very mindset I thought I had left behind.

That era of projects, Gantt charts, and rigid controls taught discipline, but it also reinforced habits I later had to unlearn. Leaders like me, the new "dinosaurs," are the ones who most need to unlearn and relearn.

Research backs this up. A study of agile transformation at Volvo Cars found that even during major reorganizations, teams maintained their maturity only when they remained structurally intact. Disrupted teams, no matter how capable, were unable to progress. My experience mirrored that finding: stability was not a nice-to-have. It was the foundation of performance.[2]

Later, I faced another moment that tested whether I had absorbed the lesson. I recall a senior leader pressing me to "borrow" two engineers from one team to meet an immovable deadline. The old reflex surfaced. On the spreadsheet, the move solved the problem neatly.

But I held the line. Instead of breaking team stability, we trimmed scope and focused on the most critical deliverables. The team delivered a leaner version on time, stayed intact, and built momentum for the next sprint.

That decision became evidence that unlearning and relearning weren't theoretical. It was practical, and it began to reshape how I led from then on.

THE SYSTEMIC TRAP: WHEN ORGANIZATIONS DON'T UNLEARN

My experience was personal, but the same dynamic plays out at the organizational level. Even when teams are ready to work differently, legacy structures often crush their progress.

That decision to hold the line and choose stability over short-term optimization was a personal victory. But as I peeled back the layers of my own leadership, I realized the problem wasn't just inside my head. It was embedded in the structures, policies, and expectations around me.

We often blame individual managers for being "old school" or "stuck in the past," ignoring the reality that they operate inside systems designed to enforce those very behaviors.

What I often saw was that even when a leader wanted to empower their team, the organizational machinery—HR policies, finance cycles, and governance models—pulled them back toward control.

I have seen this pattern repeat across the acquisitions and integrations I've observed, contributed to, and now advise on. A high-performing culture is rarely dismantled by malice; it is dismantled in the name of efficiency.

Consider the common scenario where a large organization acquires a smaller, high-performing company. The smaller company has often already undergone its transformation: modernizing its architecture and organizing small, cross-functional teams around clear ownership boundaries. These teams are not an accident of org charts. They are the product of deliberate design, often applying principles like Team Topologies to create structures with just enough autonomy to deliver outcomes.[3]

In my experience with acquisitions and integrations, I have often seen nuanced team structures set aside in favor of familiar policies before leaders take the time to understand what those structures were designed to protect.

Instead of approaching the system with curiosity, the reflex is often control: facilitation roles are consolidated, team boundaries are redrawn to meet standardized ratios, and structures built for domain alignment are reshaped to fit oversight and hierarchy. What had been optimized for autonomy and flow can quickly be pulled back toward standardization.

The consequences are predictable: teams designed around clear ownership lose the tight focus that made them fast. Progress in aligning team structures with architecture and value-stream thinking risks being unwound by structural standardization.

The lesson is not that standardization policies are wrong. They serve important purposes. But when enforced as a one-size-fits-all policy on organizations that have already matured into domain-aligned structures, the pursuit of organizational neatness can mask the erosion of autonomy, decision-making authority, and delivery speed.

This is more than a cautionary tale about acquisitions. It illustrates how legacy thinking clings to structural reflexes even when a working alternative already exists. The pattern is well documented across post-M&A research: the instinct to standardize can override the conditions that created value in the first place.

The Diagnostic Tool: CAMVID

Over time, I needed a better way to make sense of what kinds of change were actually happening. Not just in my teams, but across the organizations I was advising or learning from.

Agile ceremonies might be visible. CI/CD tools might be installed. But something still felt shallow. That's when I began developing a simple lens to assess the *depth* of change. I needed a way to spot the difference between transformation and theater.

It was late 2024 when the idea first clicked. I was editing video clips from a desk-mounted camera and paused on a single frame. Cameras capture a moment in time, but transformation doesn't happen in a still frame; it occurs in the sequence. In the edits. In the decisions that change the story.

I'm not an expert in change management. This framework wasn't built to replace that body of work. It simply became *my way of seeing and assessing change*, a personal lens I've used to reflect on what *actually* shifted during key moments in my career.

That's how I began using a shorthand framework I developed that I now call *CAMVID*, an acronym I use to assess whether change was surface-level or whether it reached the structural and cultural parts of the system that actually matter:

- Cultural
- Architectural

- Managerial
- Value-Oriented
- Individual
- Delivery Model (Software)

I started using CAMVID in conversations with peers, executive teams, and industry leaders, not to prove anything, but to surface what wasn't being talked about. Suppose we were only shifting architecture or adopting new tools, which were helpful but not transformative.

If we were changing how we lead, how we deliver, how we measure value, and how people experience work, then we were starting to undergo a transformation.

CAMVID Change Assessment

Dimension	Low-Impact Change	High-Impact Change
Cultural	Renaming roles or teams without mindset shift	Embracing autonomy, continuous learning, psychological safety
Architectural	Installing tools (CI/CD, dashboards)	Redesigning systems for modularity, speed, and ownership
Managerial	PMO-led Agile, top-down control	Flattened orgs, distributed decision-making, span of control redefined
Value-Oriented	Measuring features or output	Shifting to outcome-driven prioritization and business value alignment

Individual	Training in isolation, no role evolution	Redefining roles, enabling career growth in new delivery models
Delivery Model	New processes layered on old project model	Moving to continuous flow, product operating model, team topologies

It helps me step back and ask: What's actually changing? If we're only making surface edits, we shouldn't expect lasting outcomes. However, when we start to see shifts across all six dimensions, we know the story is changing. And in my experience, that's when the real work, and the real results, begin.

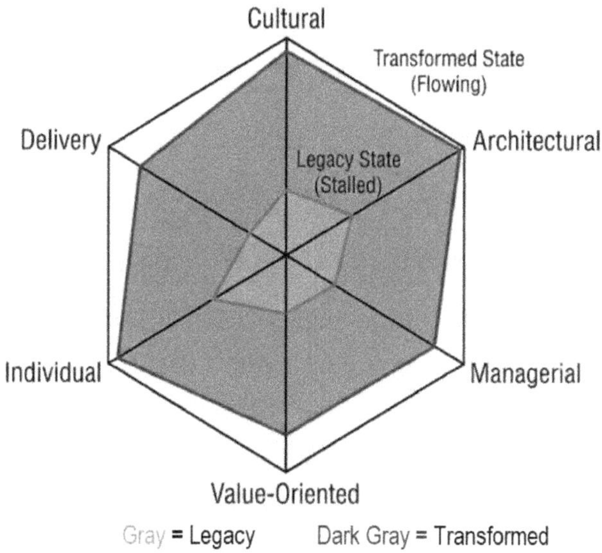

Figure 2.1: The CAMVID Maturity Diagnostic Model.

Transformation requires expansion across six critical dimensions.

Making Change Real: The Daily Discipline

Recognizing legacy thinking is one thing. Replacing it with new habits is another. Leaders often understand, in theory, that empowerment and outcomes matter more than control and output. However, the shift becomes real only when it shows up in the micro-habits of leadership: the specific words we choose, the way we run meetings, and how we react to bad news.

Here is how to operationalize unlearning across four critical dimensions:

Language: From "Resources" to Professionals

Language is the operating system of culture. Words like "resources," "headcount," or "bodies" aren't harmless shorthand; they reveal a mental model that views people as interchangeable manufacturing parts.

When a leader asks, "How many people do we need for this project?", the question often reflects an industrial assumption that more people will linearly increase output. In software engineering, Fred Brooks argued the opposite in what became known as Brooks's Law: adding people to a late software project can slow it further because the work of onboarding, coordination, and communication expands with each additional person.[4]

The Shift: Leaders who want high-performing environments must be intentional about language.

- Legacy: "I need three resources."
- Modern: "We need a team with these capabilities."
- Legacy: "Did your utilization rate drop?"
- Modern: "Did you have enough focus time?"

Changing your language forces you to change your thinking. It reminds leaders that they are leading creative professionals, not managing an inventory of parts.

LEADERSHIP PITFALL:

The Technologist Tunnel Vision

A Scrum Master at a prominent for-profit education institution shared a story with me that captured a specific flavor of legacy thinking. The organization was struggling with a large-scale SAFe implementation, and the CIO was convinced he had identified the root cause.

He believed the transformation was failing because non-engineering staff "didn't understand the developer role." In his view, the friction wasn't due to misalignment or to the operating model design. It was that people lacked technical hard skills.

His solution was blunt: a multi-week coding bootcamp for everyone in the delivery organization, regardless of role.

"It was painful to watch," the Scrum Master told me. "We had a Product Owner with deep institutional knowledge of our student systems being forced to learn coding syntax. She wasn't learning how to prioritize work or define value. She was stressing out over compiling code she would never use."

This is an example of leadership misalignment despite good intent. The technology leader confused digital transformation with turning everyone into programmers. He missed the point that high-functioning teams require different roles with different masteries. By forcing a Product Owner to be a poor coder, he prevented her from becoming a great Product Owner.

Why it matters: Digital literacy does not mean everyone types syntax. It means leaders respect the unique value each role brings to the system. When leaders try to turn everyone into engineers, they don't get better code. They get distracted roles, degraded outcomes, and a demoralized organization.

Metrics: From Surveillance to Signals

Legacy metrics are designed for surveillance: *Are people working hard?* Output-based metrics like story points, tickets closed, or hours logged are attractive because they are easy to collect. But they pull leaders back into old habits by rewarding activity over value.

The Shift: Outcome-oriented metrics take more effort, but they change the conversation from "compliance" to "impact."

- Legacy Reflex: When a deadline is missed, the reflex is to ask, "Who dropped the ball?"
- Modern Discipline: The discipline is to ask, "Where did the flow break down?"

By focusing on the work (Flow Metrics) rather than the worker (Activity Metrics), you lower the fear in the room. When fear drops, transparency rises, and you finally get the truth about where your bottlenecks are.

Role: From Gatekeeper to Enabler

In the legacy model, the manager's job is to approve decisions. They are the gatekeeper of quality and risk. This creates a bottleneck where teams wait for "permission" to move.[5]

The Shift: In a modern organization, the leader's job is to provide *context* so the team can make the decision themselves.

- *Legacy:* "Show me the plan so I can approve it."
- *Modern:* "Here is the strategic intent and the constraints; tell me how you plan to solve it."

This shift is terrifying for legacy leaders because it feels like a loss of control. But it is actually a gain in speed. Instead of approving every decision, you create the conditions (guardrails, principles, strategy) in which the right decisions happen without you.

Meetings: From Status to Solutions

The most painful symptom of legacy thinking is the "Status Meeting"—a weekly ritual where highly paid professionals go around the room reading updates from a spreadsheet. This is a control mechanism, designed to prove that the manager is "on top of things."

- *The Shift: Modern* leadership meetings are for *unblocking.*
- *The Check:* If you leave a meeting knowing *what* happened but having solved nothing, it was a legacy meeting.
- *The Goal:* Meetings should surface risks, clarify strategy, and amplify team voices. The tone leaders set signals whether they are holding power or sharing it.

The Daily Audit comes next. Making this change real requires discipline. I found it useful to ask myself three questions at the end of each day to audit my own performance:

1. *Clarity:* Did I create clarity today, or did I just add noise?
2. *Empowerment:* Did I make a decision that the team could have made?
3. *Focus:* Did I ask about "when it will be done" (output) or "what value we are delivering" (outcome)?

Some days, I slipped into old habits. Reflection helped me reset. On better days, I stayed true to the new approach, and I could see the effect in how my teams engaged and delivered.

At That Time

In my own leadership journey, I still fight the legacy reflex. Even now, when a critical deadline looms, my first instinct is often to intervene: to move a person, to demand a status update, or to ask for a "guarantee" of delivery. The difference today is that I recognize the impulse before I act on it. I use the CAMVID lens to check my own decisions: am I solving this problem Managerially (control) or Culturally (empowerment)? In many teams I worked with, the culture gradually shifted from control toward trust.

Key Takeaways

- *Legacy thinking hides in plain sight.* It lingers in language, structures, and leadership reflexes. Left unchecked, it slows innovation and disengages teams.
- *Language reveals mindset.* Referring to people as "resources" reinforces the idea that they are interchangeable parts. Changing your language is often the first step to changing your leadership.
- *Stability beats efficiency theater.* Breaking up high-performing teams to satisfy "span of control" or utilization rules destroys the very autonomy that drives speed.
- *Unlearning is a daily discipline.* Leaders replace old reflexes through repeated choices: move work rather than people, measure impact rather than activity, enable rather than control.
- *CAMVID acts as a mirror.* Use the framework to assess if your changes are cosmetic (installing tools) or transformative (changing culture and value).

Unlearning old habits is the internal work of leadership. But internal shifts alone don't ship software or generate profit. Once we have stopped the bleeding caused by legacy thinking by stopping the breaking of teams and demand for false certainty, we need to build something new in its place. We need a set of operating practices that turn this new mindset into a reliable engine for value. That is where we turn next: the specific, profitable practices that replace the old cost-center model.

3
PRACTICES THAT DRIVE PROFITABILITY

"As to methods, there may be a million and then some, but principles are few. The man who grasps principles can successfully select his own methods. The man who tries methods, ignoring principles, is sure to have trouble." —Harrington Emerson[1]

Legacy thinking explains why so many transformations stall. Leaders can unlearn outdated habits, but the harder challenge is replacing them with something better: practices that actually work. That's where this chapter begins: the specific practices that transform technology delivery from a cost center into a driver of repeatable profitability.

For decades, software delivery was managed as a cost to control. Budgets were tied to predefined projects, performance was measured in efficiency metrics, and technology teams were treated as service providers for the "real" business. While this model once seemed rational, it was designed for a slower, more predictable world.

The real conversation today is about outcomes and investment. What's the minimum viable product that delivers value within our budget so we can release sooner and get feedback faster? At a product level, how much should we invest to deliver the next enhancement customers expect, and what return are we anticipating?

The most significant improvements come when we shift from cost control to outcome alignment. This approach involves organizing teams around products or value streams, providing consistent funding for dedicated long-lived teams, and prioritizing initiatives based on their value.

The Leadership Factor in Practice Adoption

Adopting practices is one thing. Protecting them is another. Repeatable results depend not on introducing new ways of working but on leaders holding the line when pressure mounts.

Each practice below protects one of four executive stakes: profitability, risk, speed-to-learning, or organizational trust.

Long-Lived, Stable Teams

Leadership reflex: Reshuffle team members or teams to add short-term capacity or meet deadlines.

Leadership discipline: Defend long-lived, stable teams. These are small, cross-functional teams (often 7–9 people) that bring together multiple needed disciplines, not just engineers, and remain intact over time, compounding expertise and accelerating profitability.

Amazon popularized the concept of "two-pizza teams": the idea that no team should be so large that it can't be fed with two pizzas. The principle was about team size and autonomy, reducing coordination overhead, and encouraging accountability.[2]

Over time, Agile and DevOps leaders extended this principle beyond size. Research and practice across books like *Accelerate*, *Project to Product*, *Team Topologies*, and *Sooner Safer Happier* emphasized that teams should also be long-lived.[3] Constantly reforming teams erodes trust, domain knowledge, and flow. Long-lived teams build depth in their product or value stream, reduce costly handoffs, and deliver more predictably over time.

The Hidden Tax of Re-Teaming: The Tuckman Cycle

There is a well-established reason why stability creates speed, and it isn't just about camaraderie. In 1965, psychologist Bruce Tuckman defined the four phases every high-performing team must traverse:[4]

- *Forming:* The team comes together, defining roles and boundaries. (Low productivity).
- *Storming:* Friction arises as working styles collide and trust is tested. (Dip in productivity).
- *Norming:* Rhythms are established, and social contracts are settled. (Rising productivity).
- *Performing:* The team operates with high trust and implicit coordination. (Peak productivity).

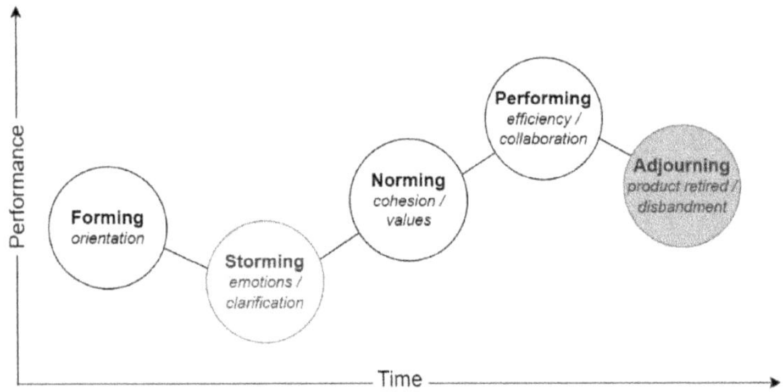

Figure 3.1: *The Tuckman Cycle and the 'Re-Teaming Tax'.*

The graph illustrates the hidden cost of instability: changing team members forces a "Performing" team back into the low-velocity "Storming" phase, destroying compounded momentum.

The legacy manager views a team member as a "resource" that can be slotted in at full capacity immediately. The reality is that every time you add or remove a person, you force the team back into the 'Storming' phase.

You aren't just moving a person; you are resetting the cycle. Long-lived teams are profitable because they stay in the *Performing* stage, compounding their velocity over time. "Fungible" teams are expensive because they are perpetually stuck in *Storming.*

The Discipline of Stability

As I described in Chapter 2, my own reflex to reshuffle teams destroyed trust. I learned the hard way that breaking long-lived, outcome-aligned teams is not a tactical adjustment; it is a strategic setback that undermines both speed and quality.

When team stability became a real discipline, the results were very different. As teams stayed together, they built cohesion, deep product expertise, and clearer ownership of outcomes. Quality improved because accountability was no longer spread thin. Delivery accelerated because knowledge didn't reset with every change.

Principle Extension: Under extreme capacity pressure, resist the urge to move people across products or value streams. Instead, move work to existing teams, or move entire teams to the work, but never fragment teams.

Why it matters: Team stability is a financial control, not a comfort preference. Every re-team delays time-to-value, increases coordination overhead, and raises defect risk, driving up cost-to-serve while slowing learning and reducing return on investment.

Continuous Funding (The Aspiration vs. Reality)

Leadership reflex: Revert to annual resets, forcing teams back into project mode.

Leadership discipline: Sustain continuous investment. It avoids stop-start waste and compounds value over time.[5]

In a traditional "Cost Center" model (Chapter 1), we fund projects. If a project is estimated to take six months, we fund it for six months. If it runs late, we have a variance conversation. In a Profitability model, we aim to fund *capacity*. We are investing in a product, through long-lived team capacity, to improve it over time and pursue specific business outcomes.

Why does this matter? Funding persistent teams avoids the waste of ramping people up and down and gives them ownership of long-term outcomes. Prioritizing value, even with rough ROI estimates, forces better trade-offs, ensuring limited capacity goes to work most likely to return investment.

REALITY CHECK:
The Gravity of Annual Budgets

I want to be transparent: in my experience, very few organizations fully achieve this "Continuous Funding" ideal. Finance departments are rarely set up for it, and the gravity of the annual budget cycle is immense.

In practice, shared agreement on ROI was not always achieved. ROI was often framed in broad terms, and the annual budgeting structure tended to pull investment decisions back toward project-based cycles.

But even imperfect attempts to link investment to outcomes shifted the conversation. The annual budget still defined the funding container, but inside it, prioritization increasingly followed rolling value rather than fixed project plans. The win wasn't changing the CFO's entire worldview; the win was shielding the teams from the stop-start nature of project financing.

Outcome-Based Metrics

Leadership reflex: Default back to counting outputs, because they're tangible and easy to report.

Leadership discipline: Demand outcomes, even if they are rough. Outcomes sharpen priorities and connect investment to profitability.

Recall the executive request for "productivity units" from Chapter 1. That request wasn't malicious; it was a desperate attempt to find a leading indicator of value. The solution isn't to refuse to measure, but to replace those output metrics with measures that actually drive *business value and* profitability.

In this book, two measures matter most: Flow and Realization.

Flow measures how efficiently value moves from idea to production. It is the operational lens.

Realization measures the business outcomes of what we deliver, such as revenue, retention, or risk reduction. It is the business lens.[6]

Shifting from cost center to profitability requires balancing both: Flow Metrics show how effectively work moves, while outcome metrics show whether that work produced results worth the investment.

Enterprise Example: In one growth-stage SaaS environment I helped lead, shifting to team-level outcomes and flow measures changed the executive conversation from activity to value. Leaders moved away from asking, "Who isn't working hard enough?" and toward questions such as, "Which teams are struggling?" and "Are we creating outcomes worth investing in?" That shift helped reframe technology from overhead to a profit-driving partner.

Mapping the Value Stream

Leadership reflex: Organize teams around short-term projects without regard to value streams.

Leadership discipline: First, align teams to value streams based on the product portfolio, domain, or sub-domain. Then enable teams to map their current and future states so they can continuously improve flow.[7]

Leaders must align teams to value streams and then support those teams in mapping their work. Alignment creates ownership; mapping reveals how value actually flows and where it gets stuck.

If you're unsure where to start with Value Stream identification, use the Value Stream Identification Template in *Appendix C: Value Stream Identification Template* to define your primary value streams and the supporting sub-value streams across your product portfolio.

Identifying your Value Streams is one approach to create clear boundaries for ownership before you begin current- and future-state mapping.

If you want to go deeper on how to adopt value stream mapping as a practical discipline rather than a one-time workshop, *Flow Engineering: From Value Stream Mapping to Effective Action* is a strong next read.

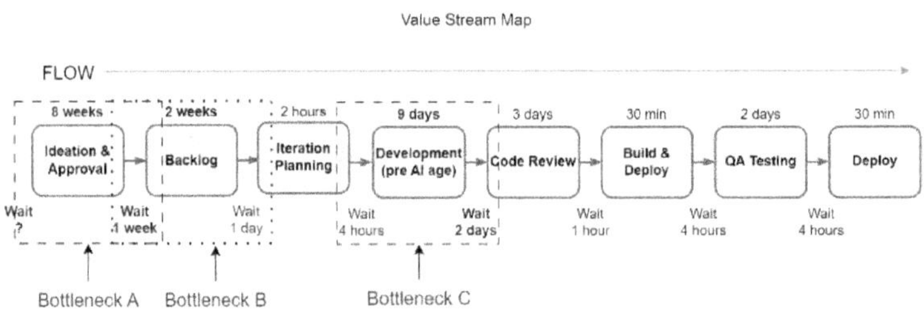

Figure 3.2: *Visualizing the Flow of Value.* A Current State

Value Stream map reveals the reality of work delivery (concept informed by Flow Engineering).[8]

The principle is that business leaders should view Value Stream Management (VSM) as a business enabler, not just a technology initiative. When embraced broadly, VSM makes work visible, improves predictability, and links delivery speed to business results.

Adoption Strategy: Adopting VSM cannot be forced everywhere at once, nor can it be owned by engineering alone. Training product and divisional leaders alongside delivery teams and gradually rolling out flow practices makes the change far more sustainable and reduces the instinctive pushback that comes with big-bang rollouts.

Why it matters: Value Stream Mapping turns delivery from a narrative into a system you can see, improve, and govern. When leaders can see queues, handoffs, and wait states, they can reduce risk and decision latency, building predictability and trust without pushing teams into heroics.

Team Experience & Developer Experience (DevEx sometimes abbreviated DX)

High-performing organizations eventually learn that productivity is a socio-technical problem: it is shaped by both the technical systems people use and the human environment in which they work. Developer Experience refers to how easy, effective, and sustainable it is for engineers to do their work within the systems, tools, and team environment around them. CI/CD, automation, and tooling are necessary—but insufficient on their own. The other half is the human system: protecting focus time, designing healthier meeting cadences, and partnering with People/HR leaders to address collaboration patterns that no pipeline can fix. That's the reason DevEx matters, and why I think of it as a second engine alongside technical capability.

A useful research lens here is the SPACE framework from Microsoft Research.[10] It reinforces the point that engineering effectiveness is multi-dimensional: satisfaction and well-being, performance, activity, communication and collaboration, and efficiency and flow. The lesson is not that leaders need to measure everything at once. Productivity cannot be collapsed into output alone without distorting what actually makes teams effective.

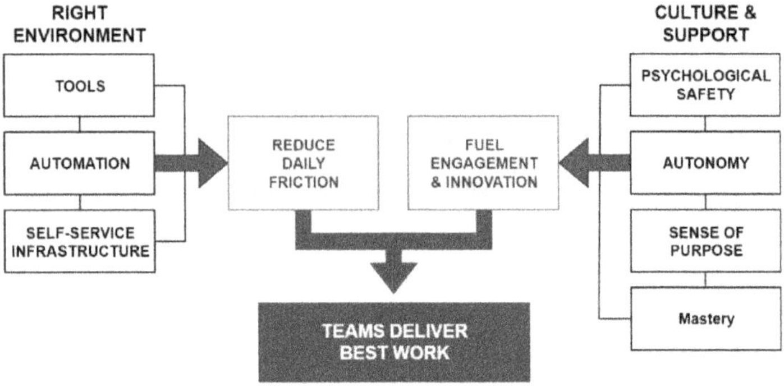

Figure 3.3: *The Dual Engine of Performance.* High-performing teams cannot exist on tools alone. They require two distinct investments: a frictionless technical environment (automation, infrastructure) and a supportive culture (psychological safety, autonomy).

Leadership reflex: Cut elements of DevEx investment first under cost pressure.

Leadership discipline: Protect DevEx and team sentiment as profitability levers and early-warning indicators of capacity, resilience, and friction.

Teams deliver their best work when they have both the right environment and the proper cultural support. Tools, automation, and self-service infrastructure reduce daily friction; psychological safety, autonomy, and a sense of purpose fuel engagement and innovation.

DevEx and sentiment are not 'soft' or merely qualitative measures. They are leading indicators of engineering effectiveness and business performance. As GitHub emphasizes, software development capacity is often the limiting factor for innovation.[9] Reducing friction and building developer confidence can improve delivery effectiveness and support better business outcomes. McKinsey research similarly links stronger developer environments and developer velocity with higher revenue growth, stronger customer satisfaction, and better business performance.[11]

The Role of AI: AI-assisted development tools represent one dimension of this environment. Like Agile, Lean, and DevOps before it, AI is not an end in itself. Left unmanaged, it can amplify the system around it, magnifying existing friction and reinforcing unhealthy patterns. But in a healthy system, one built on psychological safety, clear guardrails, and continuous learning, AI can reduce toil, shorten feedback loops, and improve team capacity. The leadership discipline is to frame AI adoption around outcomes and team support, not just speed or cost reduction.

In this kind of environment, improvements to DevEx shorten the distance between intention and impact. Teams achieve flow states more often, defects are owned and resolved faster, and delivery cadence becomes steadier.

Capacity and Focus

Leadership reflex: Say yes to everything.

Leadership discipline: Force trade-offs. Tie priorities to outcomes and capacity.

Leaders must acknowledge capacity limits and enforce trade-offs. Every new investment comes at the cost of deferring something else. At enterprise scale, the pressure to say yes to everything is constant. Product leaders often find themselves overloaded by enhancements or features that drive growth or improve customer satisfaction, shifting from outcome-driven prioritization to project scheduling.

The turning point comes when leaders find the courage to enforce explicit trade-offs: if new work enters, something else must move down. Saying no is uncomfortable, but it is the only way to protect outcomes.

Prioritization is not about saying yes to the right things, but rather about having the discipline to say no to everything else.

Prioritization as a Leadership Discipline

Capacity discipline also means a disciplined approach to prioritization. There are three essential lenses leaders can apply consistently:

1. *By Outcomes/Impact:* Anticipated ROI, revenue, retention, or customer outcomes.
2. *By Flow:* Bottlenecks and efficiency indicators.
3. *By Capacity:* The hard limit.

In a later Flowtopia presentation, I put it plainly: We will always have more ideas than capacity. Prioritize work to capacity based on Flow, outcomes, and impact. Without this discipline, work piles up, predictability disappears, and profitability suffers.

These leadership disciplines become most visible when work crosses team boundaries. Dependencies are where prioritization, capacity, architecture, and team design collide. For that reason, they deserve more than a passing mention here.

Dependencies as a Profitability Tax

Leadership reflex: Add oversight, escalation, and process every time cross-team work creates delay.

Leadership discipline: Make dependencies visible, govern them deliberately, and treat them as a system-level constraint.

Dependencies are one of the clearest places where leadership discipline either protects profitability or quietly erodes it. If you want a fast, reliable value stream, you have to treat dependencies for what they are: a recurring tax on profitability.

In modern delivery, dependencies are not rare edge cases. They are the natural consequence of building shared platforms, scaling across domains, and running cross-cutting initiatives that require multiple teams to move together. The problem is not that dependencies exist. The problem is what they do to Flow when they are unmanaged.

Here is what I mean by "tax." Every dependency introduces one or more of these costs:

- Waiting: work sits idle while someone else finishes their part.
- Coordination overhead: time is burned on alignment meetings, handoffs, and re-explaining context.
- Quality risk: pressure rises as timelines compress, and teams start trading rigor for speed.
- Rework: late changes ripple across teams, forcing retesting and re-integration.
- Trust decay: teams feel blocked, leaders feel surprised, and the system becomes more defensive.

When leaders feel that tax, the reflex is predictable: add process, add escalation, add oversight. That is how dependencies get institutionalized into bureaucracy. But the more disciplined move is different. Dependencies should be treated first as a prioritization and capacity problem, and second as an architecture problem.

In my experience, the healthiest way to manage dependencies is to stop treating them as "someone else's problem" and instead evaluate dependency work through the same prioritization logic as everything else: outcomes, Flow, and capacity. That means your roadmap, your technical debt, your operational work, and your cross-team asks all compete on the same scoreboard.

Practical Framework for Dependency Discipline

1. Unified prioritization by anticipated outcomes
2. Every request, including dependency work, must compete based on its anticipated outcome. The question is not "Who asked?" or "How loud is the request?" The question is: "What business result does this enable, and what happens if we delay it?" When teams use the same outcome lens for both roadmap work and dependency work, prioritization becomes legible and less political.
3. Capacity management that forces trade-offs
4. Dependencies become toxic when we pretend capacity is infinite. The reality is that teams are almost always operating near their limits. When a new dependency request arrives, the disciplined move is to make the trade-off explicit: if this comes in, something else moves out, slows down, or gets descoped. This is how you protect Flow without turning the system into heroics.
5. Dynamic reprioritization instead of reactive chaos.
6. Priorities change. That is not a failure; it is the nature of complex work. The failure is reprioritizing without a mechanism. Dynamic reprioritization means teams re-evaluate commitments with new information, identify what must pause, and sequence the dependency work transparently. This prevents the "everything is urgent" pattern that drives burnout and destroys predictability.

This approach also changes the role of leadership. Leaders should still engage as subject matter experts to clarify scope and risk, but the most durable decisions live closest to the teams doing the work. When leaders centralize prioritization, they often create the very delays they are trying to eliminate.

The reason this belongs in a profitability discussion is simple: unmanaged dependencies inflate time-to-value. They stretch the distance between investment and return, not because teams lack effort, but because the system spends too much time waiting and coordinating. When you treat dependencies as a visible, governable part of the value stream, you reduce that tax. Flow tightens, trust rises, and outcomes become more repeatable.

And this is where value stream mapping becomes more than a workshop. Mapping exposes where dependencies actually live in handoffs, queues, review gates, shared services, and decision bottlenecks. With this information, leaders can decide what to simplify, fund, and stop.

If you want to protect profitability, do not try to eliminate dependencies by policy. Treat them as part of the system you must lead: prioritize them honestly, fund them intentionally, and measure their impact through Flow and outcomes.

When you treat dependencies as a visible, governable tax alongside stable teams, DevEx investment, and capacity discipline, you shorten the distance between investment and return. This reflects the common thread across every practice in this chapter.

Why These Practices Drive Profitability

The common thread across these practices is that they shorten the time between investment and return. Continuous funding avoids stop–start cycles. Long-lived teams reduce handoffs and accelerate delivery. Outcome-based metrics keep the focus on value, not activity. Developer Experience and team sentiment provide early signals about capacity, resilience, and the system's ability to sustain performance over time.

Practices alone don't guarantee profitability. They only work when leaders protect them. Left unguarded, funding reverts to annual cycles, teams get reshuffled, metrics drift back to outputs, and Developer Experience budgets get cut. When that happens, profitability gains vanish as quickly as they appeared.

The leadership lesson is straightforward: profitability is not a function of tools or frameworks. It's a function of discipline. Leaders who hold the line on these practices ensure that every dollar invested moves faster through the system and produces a greater return.

What Changed and What Remained Difficult

By the end of my tenure, team stability had improved significantly; the era of "reshuffling for capacity" had largely diminished, and retention reflected that progress. Continuous funding remained a challenge. Although prioritization increasingly relied on Flow and outcomes within each fiscal year, the operating model still sat inside an annual budget cycle. Dynamic funding had not yet been solved, but there was enough buffer for teams to operate as if funding were more continuous.

LEADERSHIP PITFALL:

The Framework Shopper

In one conversation with an experienced Agile leader, I was asked for a sanity check on a change initiative that felt reactive rather than diagnostic.

Her teams had been using Scrum, but delivery was inconsistent. The Technology division was considering a broad shift to Kanban while the underlying causes of delay and friction were still not fully understood.

I asked the questions leadership should have been asking first. What is the architecture like: monolith or decoupled? How are teams aligned to products or domains? What do dependencies look like across teams? Where does work actually wait?

She paused, then sighed. "We don't know," she admitted. "We aren't discussing architecture or dependencies. Leadership thinks Scrum is too heavy and a new framework will fix the speed."

That was the gap. The Technology division was trying to solve a system problem by changing processes. It was looking for relief in ceremonies and workflow mechanics, when the deeper friction often lives in coupling, handoffs, unclear ownership, and decision latency.

Why it matters: A strategic leader does not prescribe treatment without diagnosis. If you change frameworks without understanding architecture, team design, and dependencies, you carry the same dysfunction into a new vocabulary. And the cost is not only weaker delivery. It is also the erosion of trust, learning, and improvement capacity.

Leadership Reflex vs. Leadership Discipline (Self-Assessment)

This chapter is more than a catalog of practices; it is a mirror for leadership. Each practice described here is possible. But under pressure, reflex and discipline often pull in opposite directions. The table below summarizes those trade-offs and provides a simple self-assessment for where your organization stands today.

Practice	The Legacy Reflex	The Leadership Discipline
Funding	Revert to annual project cycles; restart work every year.	Protect continuous investment; let teams compound value over time.
Stable Teams	Reshuffle for deadlines; reset learning curves.	Defend long-lived, outcome-aligned teams that accelerate profitability.
Metrics	Default to outputs (velocity, feature counts).	Insist on outcomes, even rough ones; connect investment to ROI.

Team Experience (or DevEx)	Cut budgets first under cost pressure; treat sentiment as "soft."	Treat DevEx and sentiment as profitability levers and focus on early-warning indicators.
Flow & Realization	Focus only on speed (*how fast?*) or only on outcomes (*what value?*).	Demand both; profitability requires speed and impact together.
Capacity & Focus	Treat every request as a priority; overload teams.	Enforce trade-offs; prioritize by customer or business outcomes, Flow, and capacity.

Key Takeaways

- *Profitability is about time-to-value.* Cutting scope and budgets may save in the short term. Still, the real driver of profitability is shortening the distance between investment and customer value because return follows when customers succeed.
- *Practices must be protected.* Continuous funding, long-lived teams, and outcome metrics are fragile. Leaders must defend them against the organizational reflex to revert to command-and-control.
- *Stability creates speed.* Resisting the urge to reshuffle teams allows knowledge to compound. Moving work to teams is cheaper and faster than moving people to work.
- *Outcomes anchor investment.* Even if you can't escape the annual budget cycle, use "anticipated outcomes" to prioritize work within that budget. This keeps teams focused on value rather than just "delivering features."
- *DevEx is a financial lever.* Reducing friction for developers isn't just about morale; it's about increasing the capacity of your most expensive asset.

Reflection: Pause and Consider

Before moving to the next chapter, consider your own experience during the last high-pressure delivery cycle:

- When pressure rose, did you default to reflexes (adding people, demanding status), or did you apply discipline (protecting the team, clarifying outcomes)?
- Have you demanded outcomes for every investment, or allowed vague goals to dominate because it was easier?
- Do you enforce trade-offs, or let busyness masquerade as progress?

These are not abstract questions. They determine whether your leadership produces profitability or just activity without return.

Practices alone, however, are fragile. Under pressure, they erode unless leaders reinforce them with metrics that act as steering mechanisms: signals that expose where work slows, where risk accumulates, and whether delivered work is creating meaningful outcomes.

Metrics make those conditions visible and keep the organization accountable to value rather than activity.

This is where the next part of the journey begins. Chapter 4 turns to metrics, not as dashboards or vanity reports, but as steering mechanisms that protect profitability. We will look at how to measure what actually matters to the business, giving you the evidence you need to defend the practices we just discussed.

4
Metrics That Matter: Measuring Flow, Predictability, and Outcomes

Goodhart's Law is often summarized in a simple warning: when a measure becomes a target, it ceases to be a good measure.[1]

Metrics are both a compass and a trap. Get them right, and they guide technology investments toward measurable business value. Get them wrong, and they distort behavior, hide real problems, and erode trust faster than any reorganization ever could.

I've witnessed both outcomes unfold. The difference is never the math—it's how leaders interpret and use the data.

Just as practices demand leadership discipline (Chapter 3), measurement demands interpretive discipline.

Some organizations measure only activity: features shipped, story points completed, and tickets closed. They confuse movement with progress. Others collect extensive data but fail to connect it to business results, leaving leaders unsure whether their investments are paying off.

Metrics do more than describe performance after the fact. They also shape how leaders make commitments in the face of uncertainty. That is why predictability, confidence, and Cost of Delay belong in this chapter: they are not detours from measurement, but part of how a mature metrics system informs decisions before, during, and after delivery.

This chapter explores how to build a metrics system that matters: one that improves forecasting, exposes flow constraints, protects team health, and connects delivery to business results. It acts as the defense system for your transformation, providing value to the business while protecting teams from the toxic reflexes of the past.

Why Metrics Fail (And It's Not the Math)

I recall a conversation with a senior leader during a quarterly review. They were visibly frustrated as they flipped through slides showing team velocity trends.

"Why is this team so far behind the others?" they asked. Their tone implied that someone was not working hard enough.

The room grew quiet. But when we looked more closely, the story changed. That team was deep in high-complexity architectural modernization, while others were focused on UI enhancements. The apparent gap in velocity was not a gap in effort. It reflected different kinds of work.

Yet the leader's instinct was to use the number as a scorecard, not a signal for inquiry. That is how metric misuse begins to poison culture. I have seen leadership teams compare teams to one another or use metrics to justify budget cuts, rather than using them to understand workflow and improve it. Metrics fail more often because of leadership misuse than because of flawed math.

For as long as software engineering has existed, leaders have searched for a reliable way to measure productivity. The proxies have changed, but the instinct has not. Organizations have measured lines of code, tickets closed, commits made, pull requests merged, features shipped, defects resolved, and now broader team-level signals such as throughput, flow, and stability.

The desire behind it is understandable. Leaders do not want to fly blind, and teams do not want to work blind either. But software delivery has always resisted being reduced to a single number.

The real failure mode begins when a measure stops being a signal and becomes a target. The moment leaders ask for a 10 or 15 percent increase in productivity, the system starts to bend around the number rather than around the outcome.

Teams learn which kinds of work are most likely to satisfy the target, and they naturally drift toward those activities. That can mean favoring visible output over meaningful progress, adding code where simplification would have been better, increasing commit counts without increasing value, or slicing work more thinly so throughput appears higher. The dashboard improves. The system itself may not.

I have often said that if you separate business conversations from team conversations, the numbers usually take care of themselves.

The business conversation should focus on direction, priorities, and outcomes. The team conversation should focus on bottlenecks, friction points, dependencies, quality issues, and experiments that improve flow.

When teams are given the space, support, and telemetry to inspect how work moves, where it gets stuck, and what changes might improve it, the business is far more likely to see better numbers for the right reasons.

The minute leadership turns the metric itself into the goal, it risks missing the point. A productivity target without context often drives gaming rather than improvement. In my experience, durable gains come when metrics guide reflection, learning, and system improvement, not when they pressure people to produce better-looking dashboards.

ESTIMATION AND PREDICTABILITY THEATER

Nowhere is the misuse of metrics more damaging than in the pursuit of the "perfect estimate," especially when leaders confuse a forecast with a promise.

You are sitting in a quarterly business review. The Sales VP is staring across the table, voice rising: "You said I'd have this feature by March 15th. Why don't I have it? My team has been promising customers this functionality for months."

Every software leader has lived some version of this conversation. The project ran late. The estimates were wrong. Again. The default response is predictable: we promise better planning, more detailed estimates, and stricter accountability.

But this moment teaches us something critical about the legacy mindset: velocity debates aren't really about performance. They are about predictability.

When leaders feel exposed, uncertainty becomes intolerable. Strategy shifts. Priorities compete. Stakeholders want dates to anchor their own plans. Under that pressure, the estimate stops being a forecast and starts becoming a mechanism for emotional control.

We treat the estimate as a promise, and when the complex reality of software delivery inevitably breaks that promise, we blame the team for "bad planning."

Software Is Research, Not Manufacturing

The root of this dysfunction is a category error. We manage software delivery as if it were manufacturing. Assuming that if we measure twice and cut once, the result will be predictable.

But software development is not manufacturing. It is a continuous discovery.

Every initiative involves a unique combination of people, technology, constraints, and business context. In knowledge work, unknown unknowns are often part of the terrain. You do not just execute the work; you discover important aspects of it as you go. This makes software complex, not merely complicated.[2]

When we force this emergent process into a rigid manufacturing schedule, we don't get predictability. We get *Predictability Theater*.

Teams start gaming the system to protect themselves. They pad estimates (sandbagging). They trade quality for "on-time" delivery, accumulating technical debt that will slow them down next quarter. They embrace the illusion of certainty to survive the meeting, knowing full well the plan will crumble when reality hits.

The Hidden Cost of "On Time": Interest and Principal

There is a darker consequence to Predictability Theater than just bad data. When leaders demand a fixed date for a complex problem, they aren't just buying certainty. They are buying debt.

Ward Cunningham coined the term "Technical Debt" to describe the accumulated burden of shortcuts.[3] It is a financial metaphor that every executive understands, yet few manage correctly. In Fowler's technical debt metaphor, that "interest" shows up as slower future delivery velocity.[4]

If a feature should take four days but takes six because the underlying code is a mess, that two-day difference is the interest payment.

When you force a team to commit to a date they don't believe in, you move them into *Reckless Debt.* To hit the arbitrary date, teams skip refactoring, hard-code values, and ignore documentation. They hit the target today, but they planted the seeds of unpredictability for the next quarter.

The mistake most organizations make is treating all debt as equal. It isn't. Just like a financial portfolio, we need to assess the interest rate of our technical decisions:

- *High-interest debt:* This is the toxic code that actively slows down the team or causes frequent defects. It is the credit card debt of software. Ignoring it compounds the cost of every future feature.
- *Low-interest debt:* These are minor inefficiencies that don't immediately impact delivery. Like a low-interest mortgage, we can afford to carry this debt longer to focus on growth.

The Minimum Payment Discipline

The antidote to Predictability Theater isn't just "better estimates"; it is continuous repayment.

Too many organizations treat refactoring as a lump-sum event—a massive "cleanup project" that happens only when the system is on the brink of failure. This is the equivalent of ignoring your mortgage for a year and then trying to pay it all off at once. It is costly, disruptive, and rarely successful.

A profitable engineering organization treats code health like a loan that requires minimum payments.

- *Continuous refactoring:* Every time a team touches a section of code, they make a "minimum payment" by improving it incrementally such as renaming a variable, adding a test, or simplifying logic.
- *Embedded capacity:* The discipline is not to ask for permission to refactor, but to weave it into the cost of doing business.

When leaders ignore these minimum payments in pursuit of a deadline, the principal grows. Eventually, the interest payments (the time spent fighting the code) consume 100% of the team's capacity. You didn't get predictability; you just borrowed speed from the future at a predatory rate.

Confidence Is the Contract

The fix is not to stop planning. Estimates are not the enemy. The problem begins when a forecast is treated like a guarantee before the work has earned that certainty.

Software delivery is not deterministic at the start. It is probabilistic. Early dates are not promises; they are forecasts made with limited knowledge. A healthier contract is not "Here is the date." It is "Here is the date range, here is our current confidence, and here is what must be learned to increase it."

At the beginning of a new software change, knowledge is at its lowest point. Unknowns are highest. A team may not yet know how legacy code will react, whether a third-party dependency will behave as expected, how much refactoring will be required, or whether the first implementation path will hold up under real usage.

That is why starting confidence should usually be modest. Not because the team lacks skill, but because the system has not yet revealed enough truth.

Several factors shape initial forecast confidence:

- how familiar the team is with the type of change
- how complex the change is
- how familiar the team is with the system or codebase
- the coupling and blast radius of the change
- shared-component, platform, or infrastructure impact
- how easily the work can be validated in small increments
- how much learning can occur before the target date

That last point matters. The amount of time before a target date does not, in itself, create confidence. What it creates is a learning window. If the team has time to test assumptions, ship increments, validate architecture, and reduce risk, forecast confidence can rise. If the deadline is tight and the work is novel or deeply coupled, confidence should remain lower for longer.

A strategic leader changes the planning conversation by making confidence visible.

Early in the work, the conversation might sound like this:

"Based on what is known today, we are about 50% confident in the March 15 target. The forecast depends on reducing three major technical risks: validating the integration path, confirming performance at scale, and understanding the refactoring required in the legacy workflow."

Then the team learns.

A regression test exposes a hidden dependency.

Staging behavior differs from the architecture diagram.

Customer feedback on an early increment changes the shape of the solution.

A small release validates that one of the highest-risk assumptions was correct.

As each unknown becomes known, the forecast improves.

Later, the conversation becomes:

"We have validated the architecture, retired the highest-risk dependencies, and confirmed the core workflow in production-like conditions. We are now 85 to 90% confident in the target range."

All of this assumes a reasonably stable operating environment. Every forecast assumes the system is relatively stable and that the team has enough maturity to manage disruption when it occurs. That includes the maturity of the product owner and technical leader to handle what many teams call drive-bys: unplanned asks introduced after work has already been committed.

When priorities shift mid-iteration, production issues demand support, or new requests are inserted without space to re-prioritize, the probability of hitting the target date changes.

The disciplined response is to recalibrate the forecast, not defend a date that no longer reflects reality. If teams are not given the room to reset expectations when direction changes, the problem is not the forecast. It is the operating environment, and that has to be addressed first.

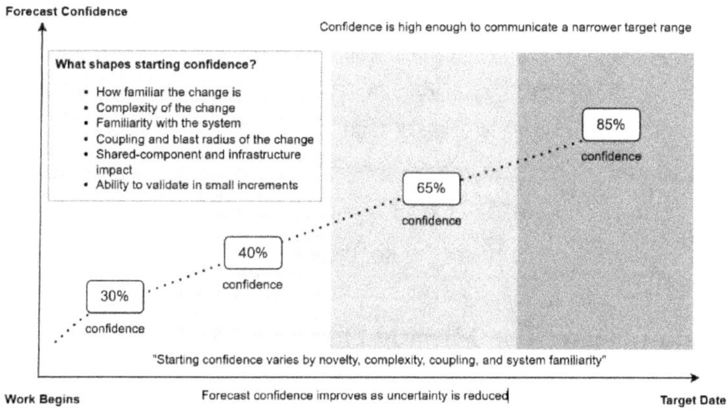

Figure 4.1: *Forecasting Software Delivery*. Forecast accuracy improves as uncertainty is reduced, allowing leaders to communicate narrower and more reliable target ranges.

That is a healthier form of predictability. The organization still plans. It still sets targets. But it stops confusing an early estimate with earned certainty.

This replaces the date as a promise with confidence in the forecast. It acknowledges that software work is discovered as much as it is executed, and that the forecast should improve as learning increases.

From Estimates to Bets

Once probabilistic forecasting is understood at the team level, the same logic should extend to portfolio decisions.

Software initiatives should be treated less like fixed-scope contracts and more like investment bets.

Bets are revisited as evidence changes.

Funding and prioritization adjust when signals change.

Spikes, pilots, and MVPs are not side work. They are mechanisms for reducing uncertainty and improving decision quality.

That is predictability without theater: smaller batches, faster learning, and forecasts that strengthen as the work earns confidence. The best response to "When will it be done?" is not false precision. It is visible progress, reduced uncertainty, and evidence from working software.

Cost of Delay: The Missing Decision Signal

A bet is only helpful if it helps leaders make trade-offs. Otherwise, it becomes a softer version of the same old plan: a roadmap with nicer language and the same hidden assumption that everything matters equally until a stakeholder yells.

There is one question that cuts through that noise:

What does it cost us to wait?

As Donald Reinertsen argues in The Principles of Product Development Flow, delay carries an economic cost, driven by the queues and wait states inside the system.[5]

That is the Cost of Delay. Not as a spreadsheet exercise. As an executive discipline.

Cost of Delay is the economic truth hiding inside every prioritization conversation. It is the price of time—revenue deferred, risk extended, opportunity surrendered, and learning postponed. And in software, time is never neutral. Even when nothing is "broken," the market keeps moving, customer expectations keep shifting, and competitors keep shipping.

This is why organizations that obsess over precision often end up making the least rational decisions. They argue about estimates while ignoring the higher cost of waiting.

THREE COMMON FORMS OF COST OF DELAY

1. Growth Delay
   The value is time-sensitive. A capability tied to retention, conversion, expansion, or market entry loses value over time. Even if the feature eventually ships, the window of opportunity may have narrowed.
2. Risk Delay
   The value is exposure reduction. Security risks, reliability gaps, regulatory weaknesses, and brittle systems all carry a cost while they remain unresolved. The longer the delay, the longer the organization stays exposed to downside.
3. Drag Delay
   The value is flow improvement. Some work doesn't create visible customer-facing features, but it removes friction that slows everything else: automation, platform stability, architectural simplification, dependency reduction. This is where "invisible work" becomes economically legible. If the system is slow, the business pays for that slowness every week.

This is also why technical debt is often misunderstood. Debt is not only a future cleanup cost—it is a compounding drag on flow. The interest shows up as longer cycle times, increased failure demand, and more energy spent navigating the system rather than improving it. In other words, debt creates Cost of Delay on everything that follows.

Cost of Delay does not require perfect math. It requires leadership clarity.

A helpful rule: avoid false precision, but refuse ambiguity.

If leaders cannot state what waiting costs. then the organization is not prioritizing. It is negotiating.

PUTTING IT INTO PRACTICE

Use this as a lightweight discipline in portfolio conversations:

- Name the delay type: growth, risk, or drag.
- Timebox the question: "If this slips by one month, what changes? If it slips by one quarter, what changes?"
- State it as a range: not a single number. Confidence matters more than precision.
- Compare delay, not effort: effort matters, but it is secondary. If two items take similar effort, should the one with the higher delay cost be done first?
- Review it as reality changes: Cost of Delay is not permanent. Markets move. Risks evolve. Strategy shifts. Update the bet.

Tip: When priorities change, document why. Dynamic prioritization only works when the reasoning is captured in the tracking system, within the initiative or related work item. Without that record, people quickly forget why the change was made and start challenging the decision without the original context.

This is decision math, not ROI theater. It doesn't pretend the future is inevitable. It simply forces the organization to acknowledge that time has a price.

When bets are framed with confidence and prioritized by the cost of waiting, the conversation finally shifts from "Are we on track?" to the only question that matters: "Did we create measurable impact?"

Predictability gets easier when flow is real

The endgame is not perfect forecasting. It's a consistent delivery.

When teams can ship safely and frequently, predictability becomes less relevant because evidence replaces promises.

That evidence changes the conversation: "We shipped value yesterday. We will ship more tomorrow. Here is what we learned, and here is our updated forecast with confidence."

One caveat matters: none of this works if the architecture and engineering practices are too fragile to ship with confidence.

If teams are drowning in heavy dependencies, technical debt, brittle systems, or poor quality, you may need to invest in architecture, capability, team design, and reduce cognitive load before you ask the organization to change how it governs work.

It is foundational to profitability.

Once leaders stop using metrics to manufacture certainty, the next maturity step is building a measurement system that captures the whole picture: system performance, team health, and business impact. Predictability is about managing flow and risk. Outcomes are about proving value.

Outputs vs. Outcomes

One of the most significant shifts in measurement maturity is moving from an outputs-first mindset to an outcomes-first mindset.

Outputs measure what we build. Outcomes measure what changes for the customer and the business as a result.

- *Outputs* are the things we produce: code changes, features released, and tickets closed. They measure activity.
- *Outcomes* are the results that outputs create: customer behavior changes, revenue increases, and cost savings. They measure value.

This distinction is reinforced by outcome-oriented frameworks such as OKRs (Objectives and Key Results)[6] and the North Star Metric[7], a product metric intended to reflect sustained customer value creation, both of which focus on changes in customer behavior and business results rather than on feature output alone.

When leaders focus on outputs, they incentivize speed over relevance. A team can deliver dozens of features that have little effect on strategic goals. Outcomes force the conversation to shift: *What result are we trying to create, and how will we know when we have achieved it?*

Anti-pattern: Setting OKRs like "Deliver five new integrations this quarter."

Better Pattern: "Increase customer retention in Segment A by 10% through new integrations."

Legacy leaders often extend this outputs-first mindset to individuals, rewarding personal throughput instead of team outcomes. But teams deliver software, not individuals. When incentives focus on individual output, people optimize for themselves: they build up an inventory of unfinished work and stop helping teammates. Collaboration erodes because everyone is measured on what they personally produce.

Leadership Insight: Outputs measure movement. Outcomes measure value.

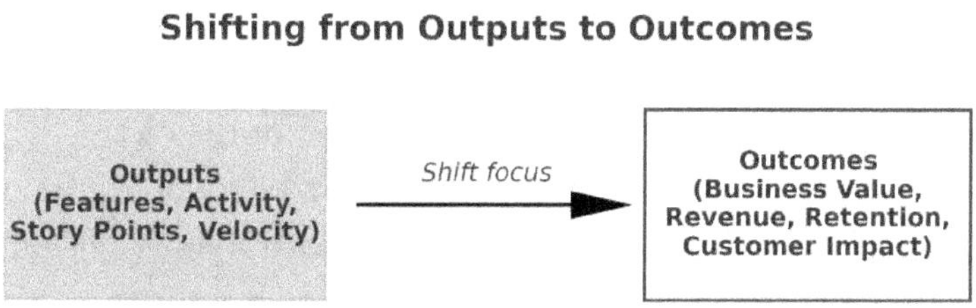

Figure 4.2: *The Shift to Value.* Moving the organization's focus from Activity (Outputs) to Business Value (Outcomes) requires a fundamental shift in measurement.

MEASURING ACROSS THREE LAYERS

To drive profitability (Chapter 3), we need to measure more than just code. Metrics operate across three layers, each telling a different part of the story.

The System Layer: Flow Metrics

Flow Metrics reveal how efficiently value moves through the delivery system. They are not about how hard people work; they are about how well the system functions.

Flow Velocity: How many flow units (countable items) are we delivering over a period? Unlike story points (subjective and team-specific), flow units are observable: features, stories, defects, service requests, or whatever your organization defines as a delivered unit of value.

A *Flow Unit* is simply the unit of work that moves through the value stream. It's the thing you can count and track from "intent" to "in customer hands." The categories vary by organization, but most Flow Units fall into four types: features (new value), defects (quality restoration), risks (security/compliance work), and debt (system improvement that protects future delivery).

Many organizations also track operational service requests as Flow Units. Once you define your Flow Units consistently, Flow Time, Flow Velocity, Flow Load, Flow Efficiency, and Flow Distribution become decision signals instead of abstract charts. You can see not just how fast work is moving, but what kind of work is consuming capacity.

We also learned a practical lesson that matters more than any metric: work has to be broken down to the smallest meaningful deliverable. In earlier years, we tried to normalize work by driving most stories, defects, risks, and technical debt items toward a small size band (for example, 3–5 points). You will never fully normalize knowledge work, and some items will still expand once you start. But over time, the discipline of small, shippable increments reduces risk, shortens feedback loops, and makes Flow visible.

The key is this: Flow Units are how we measure movement through the system. Breaking work down is how we make that movement reliable. The goal isn't perfect sizing. It's repeatable delivery, faster learning, and outcomes you can prove. We used sizing as a planning aid, not as a performance metric.

Flow Distribution is the distribution of Flow Unit investments.

Flow Efficiency: What is the ratio of active work time to wait time? In many organizations, this number is low, meaning work can spend most of its life waiting in queues rather than moving. Used alongside other Flow Metrics, it is a strong indicator of where process improvement is needed.[8]

Flow Time: How long does it take for a request to go from "approved" to "in customer hands"? This is the customer's experience of speed.

Flow Load: How much work is currently in progress? High Flow Load inevitably destroys Flow Time. It is the leading indicator of burnout and bottlenecks.[9]

Interpreting the Dashboard

Defining the metrics is the easy part. The hard part is knowing what to do when the numbers turn red. Leaders often react to bad metrics by adding pressure ("Work harder!"), But Flow Metrics require a systemic response.

Here is how to interpret the signals:

- If Flow Efficiency is low, work is sitting idle. Do not hire more engineers. Adding people to a blocked system usually increases the blockage. Instead, look for wait states: approval queues, dependency delays, or environment contention. Focus investment on removing or automating the handoffs.
- If Flow Load is high, the team is thrashing. Context switching is destroying productivity. Stop starting new work. Implement clear WIP (Work in Progress) limits and require the team to finish existing items before pulling anything new from the backlog.
- If Flow Velocity is volatile, the work may not be sized well enough, or the team may lack enough cross-functional depth to keep work moving when a key expert is unavailable. Invest in swarming practices, where the team tackles one item together, and improve how work is broken down into smaller, more consistent batches.

This is how leaders move from yelling at the scoreboard to coaching the system.

One of the most dangerous metrics is when the system looks healthy, but the people inside it feel stuck. System metrics can report strong throughput, while developers experience friction that those metrics don't capture. When that gap appears, it is almost always a signal that you are measuring the machine but not the experience of working inside it. The fix is not more dashboards. It is treating developer experience as a product, with direct feedback loops to the people doing the work.

Why it matters: Flow turns delivery from opinion into visibility. When leaders can see queues, handoffs, and wait states, they can reduce decision latency and cost-to-serve without squeezing teams into heroics.

The Team Layer: Developer Experience (DevEx)

Metrics at the team layer reveal the health of the environment. They tell you if your teams can keep improving without burning out.

We track this through Developer Experience and sentiment analysis. This is not just a "happiness score." It encompasses the friction developers face daily:

- Do they have the right tools?
- Is the documentation accurate?
- Do they have psychological safety?

Research from Microsoft's SPACE framework[10,11], later DevEx research, and McKinsey suggests that reducing friction in the developer experience improves productivity and is associated with stronger business performance, including profitability.[10]

Why it matters: DevEx is a leading indicator of whether performance is sustainable. When friction rises, quality drops and attrition risk increases, so this layer protects both margin and organizational trust.

The Business Layer: Realization

Realization connects technology investment to outcomes, and its metrics demonstrate whether the flow of work is producing measurable business impact. It is also the hardest layer to close cleanly because the signal usually arrives after the work.

We might know that we shipped Feature X and that it took 10 days, but it may take days, weeks, or even months to know whether it reduced churn, increased sales, lowered support volume, or improved retention.

In many organizations, that feedback does not sit neatly within engineering. It may live across product, analytics, finance, operations, or customer-facing teams. That helps explain why this layer is so often missing in engineering reviews.

Organizations tend to fall back on output and delivery metrics, not because they matter more, but because they are easier to see. At the Business Layer, Realization metrics answer the "CFO Question" from Chapter 1. They demonstrate that the budget allocated to the team generated a return, whether in revenue, risk reduction, or operational savings.

THE TOOLING TRAP: SEI VS. VSM

Even when leaders choose the right metrics, they can still choose the wrong visibility lens. That decision shapes what the organization can see and, just as importantly, what it remains blind to. The platform categories and examples used here reflect the market as it stood at the time of publication. Specific tools and labels will evolve, but the leadership challenge will not: choosing the visibility lens that reveals the real constraint in the system.

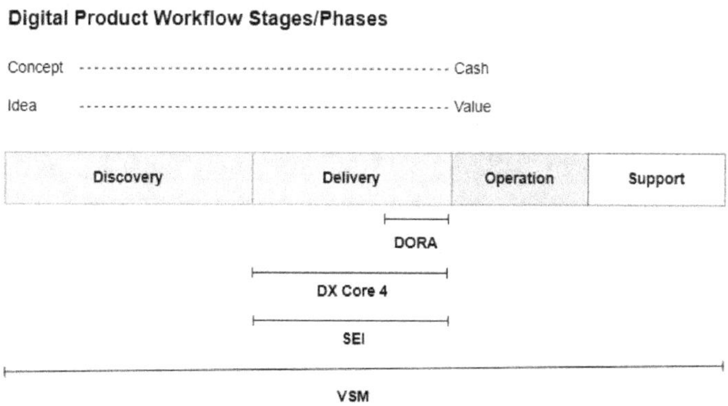

Figure 4.3: *The Scope of Measurement.* As of this publication, DORA, DX Core 4, and many Software Engineering Intelligence (SEI) platforms are primarily centered on delivery-stage performance, while Value Stream Management (VSM) platforms are designed to span a broader lifecycle from idea to value.

DORA is a powerful lens for delivery performance: how quickly you ship, how safely you ship, and how reliably you recover when things go wrong. But DORA alone can't answer the question executives ultimately care about: did this investment change anything meaningful for customers or the business? Teams can improve deployment frequency and still ship the wrong work. They can reduce lead time and still create outcomes nobody values. Treat DORA as necessary baseline visibility for the delivery stage, then pair it with flow and realization so you can connect execution to impact.

As leaders seek better metrics, they often face a confusing marketplace of tools. One of the most common points of confusion is the difference between Software Engineering Intelligence (SEI) platforms and Value Stream Management (VSM) platforms.

Both categories are valuable, but they serve different purposes. Confusing them leads to local optimization: engineering gets faster while the business remains slow.

- SEI platforms concentrate on engineering execution within the delivery stage. They typically surface metrics such as DORA[12], SPACE-informed signals[10], pull request cycle time, and related delivery analytics. These platforms are invaluable for improving day-to-day engineering workflow.
- VSM platforms focus on end-to-end value flow across the broader system. They track work from idea to realized value, exposing bottlenecks outside engineering, such as approval queues, dependency delays, or compliance reviews.

Why it matters: Each platform category creates a different visibility lens. Choose the wrong one, and you optimize the easiest stage to measure rather than the real system constraint—engineering gets faster while the business stays slow. Executives need the lens that reveals the real constraint, or they'll invest in the easiest metrics instead of the highest-leverage bottleneck.

In my experience, by the time I introduced a VSM platform, delivery was no longer the primary constraint. Teams were already deploying continuously with high automation coverage.

The remaining friction lived elsewhere: upstream in prioritization, dependencies, and decision flow. That is why I favored a Value Stream Management platform. The need was visibility across the entire value stream. An SEI platform would have given us deeper delivery-stage analytics, but the primary constraint no longer sat within delivery.

Reality Check: Don't Optimize the Wrong Thing

If your engineers are waiting on requirements for two weeks, making them code 10% faster solves nothing. Leaders must choose the lens (SEI or VSM) that reveals their actual constraint, not just the one that is easiest to measure.

When Metrics Enable the Right Decisions

While the misuse of metrics can erode culture, there are well-documented cases where leaders used them effectively to guide strategy.

McKinsey's *Developer Velocity* research found that companies in the top quartile of the Developer Velocity Index had revenue growth rates four to five times faster than bottom-quartile companies.[11]

Research documented in Accelerate identified lead time and deployment frequency as key indicators of software delivery performance. Measuring them helped organizations see how quickly work moved through the system and where delivery friction was accumulating.[12]

Why it matters: This is the executive payoff of good metrics: they replace guesswork with targeted investment. When constraints are visible, funding and attention move to the system and not to blaming teams building speed and trust at the same time.

When Metrics Turn into Mirrors

Over time, teams built trust around Flow Metrics. Teams learned to treat the data as a mirror, not a weapon: a way to see where work slowed and experiment with improvements.

That trust was tested one day when an Agile Delivery Manager noticed a single work item that had taken more than 100 days to complete.

The team's immediate reaction was fear. "We need to fix this," one developer said. They suggested resetting the ticket status from "in-progress" back to "backlog" to lower the cycle time artificially. They were worried about how the data would look to leadership.

I reminded them that the data isn't a grade. It's a signal.

We reminded them that aged work isn't failure, it's feedback. That 100+ day ticket revealed a hidden dependency that had been blocking the team for months. By making it visible, we could finally address the root cause. Hiding it would only hide the opportunity to improve.

That same fear surfaced again, this time from the middle management layer. A product manager, eager to compare team performance, began discussing Flow Metrics with a senior tech lead. He didn't mean harm, but he misunderstood the purpose. He saw the metrics as proof of performance, rather than as insight into flow.

He was comparing two of his teams' metrics when the tech lead said, "I knew it, they're using metrics against us." In an instant, psychological safety was eroded.

Those incidents taught us two simple rules:

1. Compare a team only to its past self.
2. Compare across teams only to share learning, not rank performance.

This is what separates the business and team conversations. Business discussions focus on trends and outcomes. Team discussions concentrate on bottlenecks and experiments. When those boundaries are in place, metrics serve both.

Preventing Metric Gaming

The moment you attach a performance target to a metric, people will find a way to hit the target, often by destroying the value. This is Goodhart's Law, and in engineering, it is pervasive.

I have seen organizations mandate "velocity increases" and get exactly what they asked for: teams start breaking the same work into smaller units so velocity appears to rise. A task that might once have been estimated at 5 points becomes 3 or 1-point stories. The metric improves on paper, but no real capability has changed. That is metric gaming masquerading as improvement. The velocity chart shows an upward trend to the right, but the customer sees no difference.

I have seen leaders demand "more commits per day" and get a flood of tiny, meaningless code shifts that clog the CI/CD pipeline.

How to Stop It: The only way to prevent gaming is to separate *Measurement* from *Reward.*

- *Rule 1:* Never use Flow Metrics for individual performance reviews. The moment a developer knows their promotion depends on "Cycle Time," they will stop picking up complex tickets.
- *Rule 2*: Use data for "System Reviews," not "People Reviews." Ask: "What in our *process* caused this delay?" rather than "Who caused this delay?"

When teams realize the data is used to remove the rocks in their shoes, they stop gaming the system and start using it to improve.

Google's engineering productivity research warns against relying on a single metric to measure productivity. Its central conclusion is that engineering effectiveness must be assessed through multiple outcomes, metrics, and methods.[13] Use Flow (System), SPACE-informed DevEx signals (Team), and OKRs/Realization (Business) together to get the whole truth.

This triangulation creates a balanced view. No single lens tells the story, but together they reveal whether the system is healthy, the teams are thriving, and the work is producing value.

FROM OUTPUT TO OUTCOME

How do we practically apply an outcomes-first mindset? We begin by defining the anticipated outcome before the work starts.

If a team struggles to identify the anticipated outcome that ties an initiative to a business goal, it should pause and ask why the work is being done at all. Without that connection, the effort risks becoming feature-factory output.

In delivery tools such as Atlassian's Jira, this kind of larger work item is often tracked as an Epic, but the label matters less than the discipline of defining the outcome before the work begins. Atlassian describes an epic as a large body of work that is broken into smaller pieces over time.[16]

Each major initiative should begin with a documented anticipated outcome: a measurable change in customer behavior, business performance, or operational efficiency. That outcome should connect upward to a broader business objective or key result. When the work is complete, the team returns to assess whether the intended result was actually achieved.

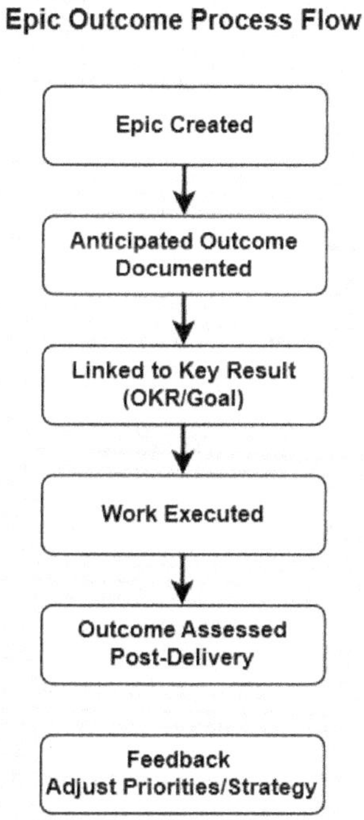

Figure 4.4: Outcome lifecycle for a major initiative. To prevent feature-factory behavior, each initiative should begin with a documented anticipated outcome and end with an assessment of actual realization.

It is easy to say "focus on outcomes," but doing it requires a different conversation. Here is a simple example of how a request can be reframed from output to outcome.

Initial request (output)

- Initiative: "Integrate the new Salesforce API."
- Why it fails: This is a task, not an outcome. A team could deliver it exactly as requested and still create little or no measurable value.

Reframing the request

- Leader: "Why are we doing this?"

- Product manager: "So the support team can see customer data."
- Leader: "Why do they need to see the data?"
- Product manager: "So they do not have to switch screens during a call."
- Leader: "And if they do not have to switch screens, what happens?"
- Product manager: "We reduce average call handling time."

Reframed request (outcome)

- Reduce average support call handling time by 15%.
- Why it works: Now the team is aligned to the result, not locked to one assumed solution. The Salesforce API may still be the right answer, but it is no longer the only answer. By defining the desired outcome first, the team has room to find the fastest, simplest, or most effective path to value.

Example outcomes

- 80% of new users complete onboarding within 7 days.
- Reduce average transaction processing time by 25 percent.

This is how the conversation shifts. Instead of funding activity, leaders define the result they want to create and then assess whether the investment actually produced it.

Where Things Stood

This experience reflects a pattern I have seen across technology organizations: progress is uneven.

Strengths: Team sentiment had been tracked for many years, providing consistent visibility into morale and friction. Team sentiment was the signal; psychological safety was the condition it helped us understand; engagement was the broader organizational outcome we were trying to support. Research shows that psychological safety and regular conversations are leading indicators of engagement.[14,15] Flow Metrics were also actively tracked, and leadership support was in place.

Gaps: At the Business Layer, OKRs were still maturing. Teams required anticipated outcomes for team-level initiatives, but upstream alignment remained inconsistent. Many functions outside engineering had not yet articulated their OKRs clearly, leaving a gap between team-level outcomes and broader business objectives. The data may have been present, but the conversations had not fully shifted.

KEY TAKEAWAYS

- Metrics are a defense system. Use Realization metrics to answer the "Cost Center" critique. Use Flow Metrics to defend against the "Work Harder" critique.
- Predictability comes from flow, not promises. Forecast with ranges based on throughput, reduce WIP, and revise confidence as you learn. Do not turn estimates into contracts.
- Triangulate your data. Use Flow Metrics at the System Layer, DevEx at the Team Layer, and Realization at the Business Layer to get the whole picture.
- Do not optimize the wrong thing. Use SEI platforms to improve delivery-stage efficiency, and VSM platforms to expose constraints in end-to-end business flow.
- Avoid the Scorecard Trap. Never compare teams against each other. Compare a team only to its past self to support continuous improvement.
- Define anticipated outcomes up front. If you cannot define the business outcome before the work begins, you are creating output, not value.

Part 1 Recap:
The Foundation

Chapter 4 closes the first section of this book. We have laid the foundation for a modern delivery organization:

What Changed in the Model

1. *Mindset (Ch 1):* We reframed technology from a Cost Center to a Strategic Partner.
2. *Leadership (Ch 2):* We confronted the Legacy Mirror and the reflexes that block empowerment.
3. *Operations (Ch 3):* We established the Practices (Funding, Teams, Flow discipline, DevEx) that drive profitability.
4. *Evidence (Ch 4):* We built the Metrics system that connects delivery to customer outcomes and steers decisions with proof, not promises.

The modern leadership shift is simple: stop defending activity and start governing outcomes. When you can show flow, explain value, and close the loop on realization, technology stops looking like overhead and starts reading like an investment portfolio.

Monday Morning Moves

- Replace "When will it be done?" with *confidence bands*, risk retirement, and updated probability as reality changes.
- Reduce thrash: *limit WIP*, expose wait states, and fix constraints instead of pushing for heroics.
- Stop ranking teams. Compare each team *to its past self* and treat metrics as steering, not surveillance.
- Make outcomes explicit: require *anticipated outcomes up front* and close the loop after release.
- Protect the system's ability to change: invest in *DevEx, automation, and self-service* to buy time back.

Proof Signals

- Forecasts become *more accurate over time* because confidence narrows as evidence increases (not because estimates get louder).
- Flow improves in ways that are hard to fake: *less waiting, less thrash, fewer "stuck" items*, more reliable movement through the system.
- Teams surface risk earlier (not later), because measurement is treated as *learning and steering*, not punishment.
- Work is increasingly described and reviewed in terms of *outcomes achieved*, not just outputs shipped.
- DevEx friction decreases and "invisible work" is made visible—because the system is being treated as a product to improve, not a machine to pressure.

The Pivot Ahead

Metrics can expose whether practices are working, but they cannot protect those practices on their own. Without the right structures around them, annual budgets pull teams back into projects, and leadership reflexes fragment long-lived groups.

Mindset and measurement set the direction; structure makes it real. In Part 2, we will examine the operating models and organizational designs that determine whether transformation thrives or stalls.

PART 2: THE TRANSFORMATION PROCESS

Fast flow happens when friction fades.

Mindset and measurement form the foundation of modern delivery, but they are fragile without the right structures around them.

Many transformations stall not because teams lack skill or will, but because operating models, funding cycles, and leadership reflexes quietly pull them back into project-era habits.

In Part 1, we reframed technology as a strategic partner, confronted the legacy thinking that erodes progress, and established the metrics that shift our focus from outputs to outcomes. These lessons answered *why* transformation is necessary and *what* it requires.

Now we turn to the more complex work: *Execution.*

This section moves from ideas to infrastructure—from how leaders think to how they build the systems that make those beliefs durable. We will examine the core systems that determine whether modern practices take root:

- *Structure:* Shifting from project-based funding to product operating models that protect flow.
- *Behavior:* Evolving leadership reflexes from command-and-control to disciplined decision-making and clear priorities.
- *Navigation:* Overcoming the silos and politics that prevent strategic alignment.
- *Culture:* Investing in the trust and developer experience that serves as the engine of performance.

The goal of this section is not to provide a rigid blueprint, but to show how structure and leadership decisions determine whether the Product Operating Model, Agile, DevOps, and Value Stream practices take root or unravel under pressure.

Transformation is no longer just about what you believe; it is about the systems you build to protect that belief.

Part 2 begins with Chapter 5, where we confront the structural shift from projects to products to value streams.

5
REDEFINING PRODUCT OPERATING MODELS

"In no other organizational structure is the ratio between output and effort needed for internal cohesion as unfavorable as in the systems structure." —Peter F. Drucker[1]

Why Structure Matters

The transformation looked perfect on paper: cross-functional teams, Flow Metrics, outcome-based planning—all the right pieces were in place.

When ownership and management responsibility shift—whether through acquisition, reorganization, or leadership transitions—team structures, governance, and decision rights often change with them. Practices that had worked well can become harder to sustain. This is why structure matters: operating models must be maintained intentionally, or they decay under pressure.

The first four chapters of this book explored how mindset and measurement shape transformation. We reframed technology as a strategic partner, built the practices that drive profitability, and learned how to measure what truly matters. But I've observed within the industry that practices erode under organizational pressure if the structure doesn't protect them. That is a hard truth: none of it endures without structure.

The best intentions collapse if the organization itself pulls in the opposite direction. If you tell a team to "be agile" but fund them annually, manage them via Gantt charts, and disassemble them once the project code is "complete," you aren't transforming anything. You are just stressing out your people.

Now we must move from ideas to infrastructure.

How do leaders think and measure? How do they build the systems that make those beliefs real?

This chapter asks the structural question: how do we organize, fund, and govern work so that practices like long-lived teams, Flow Metrics, and outcome-based measurement don't erode under pressure?

Understanding the System Before Improving It

When organizations try to improve engineering performance, they often start with practices. Leaders introduce Agile ceremonies, invest in DevOps tooling, or implement new delivery dashboards. Sometimes those efforts succeed. Often, they stall.

The reason is rarely the practices themselves. More often, improvement begins before the organization has clearly defined the system it is trying to improve.

Before leaders can optimize delivery, they must first understand how value moves through the organization. What problems are customers trying to solve? Which products address those needs? Which value streams deliver those products? And how are teams, architecture, and funding aligned to support that flow?

Without first defining the system at this level, improvement efforts become disconnected. Teams adopt practices locally, but the overall system continues to struggle with handoffs, queues, and conflicting priorities.

Understanding the system is the first leadership responsibility.

That lesson became clearer to me in hindsight through my own experience. In one technology division I helped lead, our teams pushed improvements outward from the middle of the system. We adopted Agile ways of working, invested in DevOps practices and automation, and introduced measures that helped us see delivery more clearly. The teams made meaningful progress. Flow improved, delivery became more reliable, and the transformation produced real results.

At the same time, we were still learning the larger system while trying to change it.

Looking back, I believe the transformation could have moved faster and with less friction if the broader structure had been clarified earlier: the products, the value streams, the architectural boundaries, and the operating model supporting them.

At the same time, most organizations do not start with a perfectly defined system. Improvements often begin within a single team, department, or technology group. That is normal. The important step is expanding the conversation from local improvements to the broader system that surrounds them.

The best time to clarify the system is at the beginning of a transformation. The next best time is now.

That does not invalidate the progress. It sharpens the lesson.

Modern engineering disciplines offer many useful lenses for examining that system.

Product operating models clarify how organizations structure teams around outcomes. Domain thinking helps identify the business capabilities technology must support. Value Stream Mapping exposes the path work takes from idea to operation. Practices such as Agile and DevOps influence how teams plan, build, and release software.

None of these models is universally correct, and none is inherently wrong. Each offers a different way to influence the system.

The responsibility of leadership is not to defend a particular framework but to choose the disciplines that best address the constraints visible in the organization.

The goal is not methodological purity. The goal is to improve the system that delivers value.

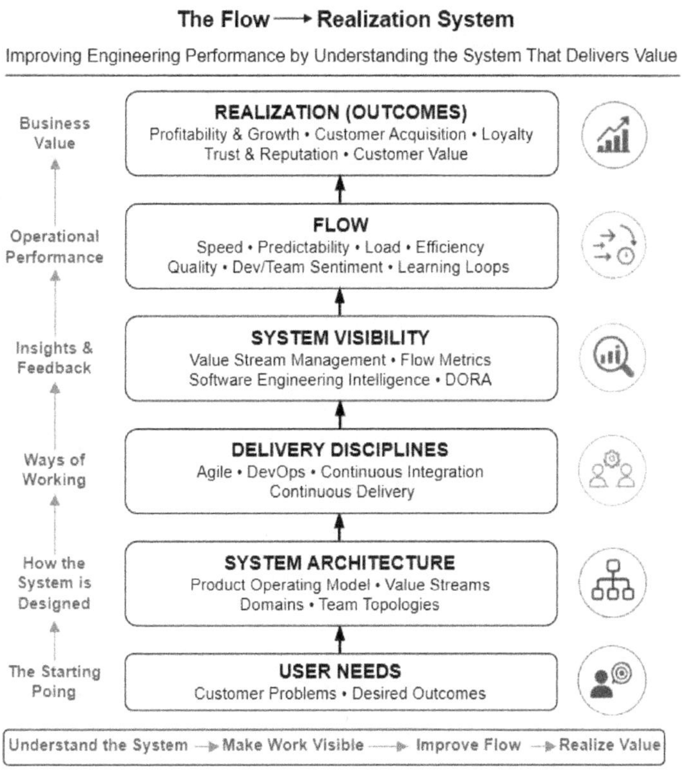

Figure 5.1: The Flow Realization System

Improving engineering performance begins with understanding the system that delivers value. Leaders clarify user needs, define the structure of that system, make work visible through flow metrics, and apply disciplines that improve the flow of value through the organization.

A more detailed version of this system diagram appears later in Chapter 12.

Seen this way, practices that sometimes look like competing schools of thought—Agile, DevOps, product operating models, Domain-Driven Design, and Value Stream Management—are better understood as tools for shaping different parts of the same system.

Some disciplines help clarify the problem space, such as understanding user needs and customer journeys. Others help define the system's structure, such as domain boundaries and value streams. Still others improve how work flows, influencing delivery practices, architecture, and team interactions.

Measurement plays a distinct role. Metrics such as Flow Metrics or DORA indicators do not improve delivery on their own. Their purpose is to make the system's behavior visible so leaders can identify where improvement is needed.

Once the system is visible, disciplines can be deliberately applied to address the constraints that arise.

This perspective helps explain why many transformation efforts struggle.

Organizations often begin improvement in the middle of the system. They introduce Agile ceremonies, invest in DevOps tooling, or establish productivity dashboards before clarifying products, value streams, or architectural boundaries. When the underlying structure remains unclear, these practices can generate activity without producing lasting system-level improvement.

High-performing organizations take the opposite approach. They begin by understanding the system, then make the flow of work visible, and only then apply disciplines to improve it.

Over time, those improvements change the signals leaders observe: work moves faster, dependencies shrink, learning cycles accelerate, and delivery becomes more predictable.

But faster delivery is not the ultimate goal.

The goal is to realize greater value from technology investments.

With this broader context in mind, we can now return to the structural question that shapes how work actually moves through the organization.

Why do so many companies claim to operate with products and value streams while still behaving like project organizations?

The answer begins with the persistence of the project model.

The Trap of the Project Model

For decades, technology work has been managed through *projects*: annual budgets tied to scope, teams formed and disbanded based on resource availability, and success measured by deadlines and deliverables.

This model made sense in manufacturing or construction, where the cost of change is high and "done" is a physical reality. When you build a bridge, you don't want to iterate on the foundation after the concrete has set. But software is never "done." Products evolve, markets shift, and customer expectations reset overnight.

Despite this reality, many organizations claim to be product-led but still cling to project reflexes. They rename teams as "products," add Agile ceremonies, and create roadmaps in tools like *Jira* or *Azure DevOps*. Yet, underneath the surface, the machinery remains unchanged: budgets remain annual, teams are redeployed for "capacity," and success is still measured by the scope delivered rather than the value realized.

The result is transformation theater: new labels on old structures.

The Psychological Comfort of the "Green" Status Report

Why is the project model so hard to kill? It isn't just inertia; it is psychological comfort. Project management offers the illusion of certainty in an uncertain world.

"Green" is not a signal. It's a story we tell ourselves to feel in control.

In complex work, the most reliable status update isn't a color. It's evidence: a working slice shipped, a risk retired, a learning captured, a forecast updated with confidence.

Progress conversations still matter. I'm not opposed to discussing initiative health, percent complete, or whether we are trending toward the target date. The shift is what those conversations are anchored to.

In knowledge work, scope evolves as teams learn. New stories emerge, technical debt surfaces, risks are uncovered, and timelines shift. That is why status reporting cannot stop at a red, yellow, or green signal. It has to show what changed, what was learned, how confidence has moved, and what that means for the forecast. Progress is not the color of the slide; it is the quality of the evidence behind the next decision.

The shift is subtle but decisive: stop managing by declared certainty and start managing by demonstrated progress. That's how you replace predictability theater with real predictability: flow, feedback, and decisions that adjust as reality changes.

The "Red/Amber/Green" (RAG) status report is a seduction for executives. It reduces the terrifying complexity of software engineering—the unknown dependencies, the shifting API contracts, the emergent bugs—into a primary color. A "Green" project promises that we are in control. It implies that if we execute the plan, the result is guaranteed.

Shifting to a Product Operating Model removes that security blanket. When we move to value streams, we stop asking "Is the project on time?" and start asking "Is the value flowing?" and "Are we solving the customer's problem?"

The answers to those questions are rarely a simple "Green." They are nuanced. They require leaders to engage with ambiguity rather than delegating it. The consequence of avoiding this shift is always the same: Agile practices collapse because the structure pulling the strings (finance, PMO, and executive reporting) is still designed for projects.

Table 5.1—Contrasting Models of Work

Dimension	Project Thinking	Product Thinking	Value Stream Thinking
Funding	Annual, tied to scope and deliverables	Continuous, tied to teams and outcomes (aspirational)	Continuous across domains, aligned to flow
Team Structure	Temporary, disbanded post-project	Persistent teams own outcomes	Teams aligned end-to-end across the value stream
Success Measure	On-time, on-budget delivery	Customer and business outcomes	Flow efficiency + outcome realization
Leadership Reflex	Move people to work	Give teams autonomy within domains	Protect flow, invest in reducing constraints

Principles of a Product Operating Model

The principles of product-oriented structures have been well documented in *Project to Product*, *Team Topologies*, and McKinsey's research on product-based operating models.[2] Taken together, they point to a few principles leaders need to internalize:

1. Long-lived, cross-functional teams compound knowledge. As Chapter 3 showed, these are non-negotiable for profitability. When a team has the skills and capabilities needed to implement changes and remains together over time, it learns the code, the domain, and how to work effectively together. That accumulated trust and shared context allow it to move faster with lower risk. The challenge is protecting that stability from the constant reshuffling common in traditional PMOs.
2. Outcome-driven planning. As Marty Cagan and Felipe Castro argue in *Outcomes Are Hard*, many alignment problems are really problems of poorly defined outcomes. If teams cannot see the outcomes they are driving toward, structure alone will not save them.[3]

3. Flow-aligned structures. This means minimizing handoffs. If a team has to file a ticket with a database group, then wait for a firewall change from a security group, and then wait for a deployment from a release group, the system has been designed for queues, not flow.
4. Continuous investment. This replaces stop-start cycles. For many organizations, this remains aspirational. Annual budgeting continues to reset priorities, even as leaders try to keep teams stable despite the fiscal calendar.

Why it matters: the operating model is a financial strategy in disguise. When leaders fund stable teams against durable outcomes, they reduce coordination cost, shorten time-to-value, and stop paying the recurring "project tax" of churn, re-teaming, and rework.

BEYOND DOGMA:
A Context-First Approach

In shifting to this model, I've learned to be wary of dogmatic interpretations. Frameworks like Team Topologies,[4] Marty Cagan's Product Operating Model,[5] and Domain-Driven Design (DDD)[6] all offer useful guidance. Team Topologies helps leaders think about team interactions and cognitive load. Cagan's model centers empowered teams around outcomes. DDD helps define software boundaries around real business domains. But none stand on their own, and none override the realities of your architecture or system constraints.

Matthew Skelton makes an excellent point: Not everything should be labeled a product, and not every product deserves its own team.[7]

I agree, but I extend the reasoning one layer deeper. For me, the starting point is always the system: the architecture, the domain boundaries, and the flow of value. When those become clear, team design stops being a matter of preference and starts becoming a matter of fit.

When I evaluate team design, I look at six things: the architecture and where the hard boundaries already exist; the domain boundaries and separations of concern; the value streams and how work actually moves; the cognitive load a team can realistically carry; the depth of specialized expertise the work requires; and the path by which customer value actually reaches the user.

These are the forces that tell you which tool to use. Cagan's empowered teams work best when the architecture supports autonomy and the team has the cross-functional skills needed to design, build, test, release, and support its work. *Team Topologies* is especially useful when multiple products share deeper system boundaries and team interaction becomes part of the design problem. Domain-Driven Design helps reveal where the boundaries truly are, not where leaders wish they were. Value Stream Management shows how work flows across those boundaries and where it gets stuck.

My stance is simple: organize teams around the natural boundaries of the system and the flow of value, not around arbitrary product labels or org-chart symmetry. The balancing act is always the same: architecture, cognitive load, expertise, and sustained demand. When those fit together, teams can stay intact long enough for ownership, trust, and domain depth to compound. When they do not, leaders end up funding coordination costs rather than value delivery.

This clarifies a frequent misunderstanding of the Product Operating Model. My understanding of Marty Cagan's message in *Transformed* is that the goal is not larger teams for their own sake, but complete teams capable of solving problems end to end.[8] In that sense, the goal is not size; it is completeness.

We must avoid dividing teams into narrow technical slices—like a "UI Team" or "Database Team"—that cannot deliver value without handoffs to three other groups. When a team owns only a fragment of the flow, ownership becomes diluted and dependencies skyrocket.

High-performing teams are built through stability. Long-lived cross-functional teams aligned to a product or service build deep domain expertise, shared context, and trust over time. Constant reshuffling may look efficient on a spreadsheet, but it creates hidden coordination costs that slow delivery and increase risk.

That stability is what makes the next step possible. In my experience, when we refactored tangled systems and clarified domain boundaries, our teams often became **smaller**, not larger. But crucially, they became **whole**. They owned a complete slice of value that they could deliver with minimal coordination.

The Context-First Heuristic:

- **The 1:1 Model:** If a subdomain or product has enough architectural independence, sustained demand, and cognitive load, then a 1:1 team-to-product model is appropriate. This is where Cagan's "empowered product team" fits perfectly.

- **The Portfolio Model:** If multiple smaller products or subdomains share architecture, dependencies, or customer value, grouping them into a "portfolio team" reduces fragmentation and improves flow. This is where Skelton's value-stream logic applies.

The core principle is that team design is a balancing act between architecture, cognitive load, and sustained demand—not just the number of "products" in your catalog.

In practice, autonomy is not a slogan. Without clear ownership and fast decision paths, teams end up carrying responsibility without the authority or support to deliver.

Case Highlight — Netflix's "Full Cycle" Model

Netflix offers a well-known example of build-and-run ownership in practice. In its engineering writing, Netflix described a knowledgeable and effective across the software lifecycle, not engineers who simply hand work to another team once code is written.[9]

Netflix also paired that ownership model with strong internal tooling. In its earlier engineering writing on how code is built, Netflix described a standardized delivery path and tooling ecosystem that functioned as a paved road. Public summaries of that post describe a code delivery process built around shared tooling and deployment infrastructure, and the post itself emphasizes that tools gain adoption when they are compelling, reduce cognitive load, and are better supported than alternatives.[10]

Viewed through a Team Topologies lens, this is a strong example of how end-to-end ownership works best when team autonomy is reinforced by platform capabilities rather than burdened by unnecessary infrastructure complexity.

Practical takeaway: In one approach I have seen work well, organizations do not require every stream-aligned team to own every infrastructure concern end-to-end. Instead, they combine strong product-team ownership with clear platform support and stable interfaces. At scale, platform support works best when it behaves like a product rather than a concierge service.

If every request becomes bespoke, the platform team becomes the next bottleneck. Self-service works best when offerings are clear, interfaces are stable, guardrails are built in, and change happens on a predictable cadence.

That is why this discussion does not end with people and process. The operating model also determines AI readiness.

Future-Proofing: The AI Prerequisite

There is another reason this model matters: AI will not fix a system that is already unclear, fragmented, or overloaded. It will amplify it. Organizations with clear ownership, manageable cognitive load, stable interfaces, and strong platform support are far better positioned to turn AI into useful leverage rather than additional noise.

Many organizations are rushing to deploy AI development tools and agents (such as *GitHub Copilot, Codex,* or *Claude Code*), hoping for a productivity boost. But applying AI to a fragmented system only accelerates the chaos.

- *Without a value stream:* AI makes coding faster, but it doesn't solve the bottlenecks in ideation, funding, requirements, or release. You reach the red light faster by optimizing locally at the cost of total confusion.
- *With a value stream:* You have a clear map of the Flow. You can point AI agents at specific friction points—automated testing, requirements parsing, or anomaly detection—across the *entire* lifecycle, not just the IDE.

Structure is no longer just about people. It is increasingly the operating interface for AI-enabled work.

Story: A Success Story with One Critical Blind Spot

To understand how these principles play out in reality, I look back at my experience at Parchment. It was a success story, but one with a specific blind spot that many technology leaders will recognize.

The technology organization led much of the transformation. Automation expanded, teams shifted to cross-functional, self-managed, long-lived structures, DevOps practices took hold, and Flow became visible.

In the early stages, a strong technology leader guided, supported, and protected that work. He served as the executive voice, helping shift mindsets away from static, twice-a-year delivery models in a rigid industry and toward continuous delivery.

But there was a catch: product leadership lagged.

Product managers often learned Agile ways of working reactively, through the delivery guardrails already in place, rather than through deliberate training or leadership development. Some questioned the need for roles like the Scrum Master or Agile Delivery Leader. Others saw themselves as team bosses rather than collaborators. Some fell into the opposite trap, acting as order-takers for senior stakeholders and simply passing down feature requests instead of helping teams solve problems.

None of this was malicious. It was the natural outcome of not investing enough in education around modern practices. Product managers were not given structured training. Cross-functional team members did not share a common library of mental models. Too much depended on Agile Delivery Managers and Scrum Masters coaching people informally in the hallways.

This is not criticism. In my view, many of the decisions product managers made at Parchment supported the company's long-term growth and success. But my observation is that because the operating model understanding was not fully shared across Technology and Product, teams experienced more friction than necessary.

The lesson: the stronger approach is to bring product, engineering, and leadership onto a shared learning baseline and a common understanding of the operating model. When people start with the same language and the same foundations, transformation takes hold faster and runs deeper.

Context Matters

These principles do not play out identically in every environment.

In heavily regulated industries such as defense, government, or finance, some degree of command-and-control is unavoidable. Compliance, safety, and risk management require it. You cannot "move fast and break things" when you are managing air traffic control software or swift banking transactions. On the other end of the spectrum, small digital startups often operate with near-total fluidity.

But even in highly constrained environments, leadership can evolve. Leaders may not eliminate controls, but they can change *how* those controls are experienced: shifting from gatekeeping to coaching, from punitive reviews to learning loops, and from siloed decisions to cross-functional collaboration.

The point is not that every organization can be "free-flowing," but that leadership has a choice in how to design structures that protect Flow and outcomes within their realities.

Operating Model Patterns Across Contexts

Not every organization can, or should, adopt the same operating model. What matters is understanding the constraints you operate under.

- *Heavily governed environments (defense, government, finance):* The mistake here is assuming controls must be centralized in ways that fragment flow. Leaders can retain control and redesign how they are experienced: fewer gates, more precise criteria, and more frequent lightweight reviews.
- *Scale-ups and mid-market enterprises:* These companies often have the most to gain from adopting value streams. They are big enough for silos to slow them down and small enough to move with intent. Sequencing matters here: start with vision, then team alignment, and only then refine governance.
- *Startups and product teams with high uncertainty:* It's tempting to say "we're too small to need structure." In practice, a minimal operating model helps small teams avoid local optima. Naming a value stream prevents flailing.

Across all contexts, leaders still face a choice: do we design for *Flow and learning* within our constraints, or do we let the constraints dictate our design?

Leadership Mindset Shapes Architecture and Teams

Team structures are not neutral. They are a mirror of leadership mindsets, past and present. Over the years, I've seen how closely system architecture and organizational design go hand in hand. Every organizational chart is a reflection—a snapshot of what leadership really values, not what it says it values.

Early in our journey, we transitioned from functional manager-led teams to small, cross-functional, self-managed teams. At the same time, our architecture transitioned from a monolith to microservices, with smaller domains and subdomains. Those microservices weren't just "smaller services"—they formed a *distributed system*, where each domain owned its own logic and data.

Autonomy in the system supported autonomy in the teams, and leadership's mindset enabled that shift.

CONCEPT IN BRIEF:
Domains, Subdomains, and Microservices

To structure teams correctly, we must agree on what they actually own. The terms "domain" and "microservice" are often used interchangeably, but they are distinct concepts that leadership must disentangle.

- **The Domain (The Problem):** This is a specific area of business logic or capability (e.g., *Payments*, *User Identity*, or *Inventory Management*). It is defined by the business need, not the code.
- **The Subdomain (The Slice):** Complex domains are often too big for one team. We slice them into subdomains (e.g., within *Payments*, you might have *Invoicing* and *Credit Card Processing*).
- **The Microservice (The Solution):** This is the technical artifact—the actual code and database—that solves the problem of the subdomain.

The Leadership Trap:

Many organizations create teams by layer (Database Team, UI Team) rather than by domain. In a Product Operating Model, the team boundaries should match the Domain boundaries. When a team owns the Invoicing Subdomain end to end (including its specific microservices), they can deliver value independently.

We refactored and redesigned our architecture to adopt a distributed microservices architecture. This allowed us to form smaller, distributed teams focused on specific business domains.

Conway's Law in Plain Sight

As Melvin Conway observed in 1968, organizations design systems that mirror their communication structures.[11] Team structures don't just mirror architecture; over time, they *become* the architecture. Leaders who say "we organize one way but architect another" are fighting physics.

Suppose you organize your people into a Database Team, a UI Team, and a Business Logic Team. In that case, you will inevitably build a three-tier architecture with rigid boundaries and queues between them. If you want a microservices architecture that flows, you can organize your people into stream-aligned teams that possess all the skills (DB, UI, Logic) to deliver independently.

The "Inverse Conway Maneuver"

A practical way to bring structure and architecture back into alignment is what the authors of *Team Topologies* call the "Inverse Conway Maneuver."[12] Instead of letting the architecture dictate the team, you design the team structure to *force* the desired architecture.

How to do this practically:

1. *Map one value stream from idea to operation.* Name the architectural streams or dependencies it crosses.
2. *Map teams to streams.* If a single team spans many value streams, expect slow Flow and hidden queues.
3. *Consolidate decision rights* where the code and runtime responsibility actually live. Don't separate design, build, and run across three committees if one team owns the service from end to end.
4. *Move work, not people.* This is the golden rule of the Product Operating Model. When demand spikes for a specific product, do not break up a high-performing team to "lend" developers to that area. Instead, route the work to the team that owns that stream. If that team is overloaded, route adjacent work to other teams or invest in a new cross-functional, long-lived team. Do not slice the humans.

5. *Tune interfaces, then teams.* If streams are noisy (characterized by tight coupling and shared mutable state), fix the software interfaces first; reorganizing teams before fixing the interfaces moves the pain.

When leadership leans into autonomy, you see modular code and stream-aligned teams. When leadership leans back into control, both drift toward a command-and-control approach—organizationally and technically.

Integration and Leadership Philosophies

Two organizations can both be "right" and still struggle to integrate.

In one integration I experienced, we were bringing together teams that had evolved under very different operating models. One side leaned toward centralized coordination and shared components; the other relied on small, cross-functional teams aligned with clear product domains and local decision-making. Both approaches had produced results in their own context.

The friction showed up quickly—not as a debate about talent, but as a mismatch in assumptions: how decisions get made, where accountability lives, how work is prioritized, and what "good" looks like. The domain-aligned autonomy created speed, but it also made the model harder to reconcile inside a more centralized governance structure. And on the other side, leaders had confidence in a playbook that had worked for years—so the default instinct was to preserve it.

That posture wasn't dismissal. It reflected a different leadership philosophy.

One model optimized for consolidation: larger manager-led groups, standardized practices, and centralized decision paths designed to drive consistency at scale. The other optimized for distribution: leadership and ownership spread across product, technical, agile, and quality roles, with teams designed to be stable, domain-aligned, and locally empowered.

The real challenge isn't whose model is "better." It's recognizing that leadership philosophy leaves fingerprints on architecture, team design, and culture.

Integrating organizations in a healthy way requires context, discipline, patience, and strong judgment—because you're not just merging org charts; you're reconciling incentives, decision rights, and how work actually flows.

Over time, we learned what I call the Price of Alignment. In the pursuit of a clean, unified org chart, they unintentionally reduced the system fit that had made the teams fast. They traded domain ownership for organizational symmetry—and didn't fully appreciate that when you destabilize team boundaries and decision rights, you often destabilize the architecture right along with it.

When two leadership philosophies meet, compromise isn't the hard part. Making assumptions, authority, and tradeoffs explicit is.

Case In Point: When Two Operating Models Collide

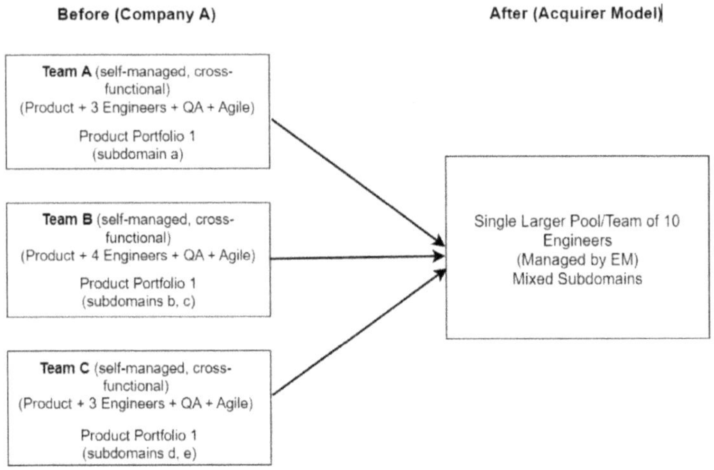

Figure 5.2: The Price of Alignment. Consolidating teams for organizational symmetry often breaks the architectural flow, illustrating Conway's Law in reverse.

Across acquisitions, integrations, and conversations with peers navigating similar transitions, a familiar pattern emerges.

An organization spends years shaping its delivery system so that team structure, ownership, and architecture reinforce one another. Small, cross-functional teams carry clear domain responsibility. Leadership roles support healthy delivery rather than command-and-control oversight. The system works because the structure fits the work.

Then a larger enterprise introduces a different operating philosophy, one that values broader spans of control, more centralized governance, and greater structural consistency across the organization. The instinct is understandable: simplify reporting lines, standardize team design, and make the whole system look easier to manage.

On paper, it often looks cleaner. In practice, it can create friction. The close fit between team boundaries and architectural boundaries begins to loosen. Managers inherit a broader scope without equivalent support. Ownership becomes less distinct. Teams that once moved with confidence within clear domains now operate in more ambiguous contexts and with greater cognitive load.

That does not mean standardization is wrong. It means standardization has limits. When structural consistency is applied without regard for how the delivery system actually works, efficiency on the org chart can come at the expense of effectiveness in the work.

Research supports the concern. Post-acquisition integration can destroy value when leaders dismantle the operating conditions that made the acquired organization effective in the first place. Team Topologies makes a related point: when teams absorb more cognitive load than they can reasonably handle, quality and speed both decline.

The lesson is not that broader spans of control or more consistent governance are inherently flawed. In many settings, they are useful. But in engineering organizations built around domain ownership, manageable cognitive load, and clear decision boundaries, imposing a more generalized model can weaken the very conditions that made delivery effective.

When two leadership philosophies meet, compromise is rarely the hardest part. The harder work is making assumptions, authority, ownership, and tradeoffs explicit.

Funding: The Stubborn Barrier

We saw in Chapter 2 how annual budgets quietly anchor organizations back to project thinking. The structural impact becomes sharper here: when investment does not preserve stable product ownership through long-lived teams, organizations fall back into stop-start cycles that erode continuity and learning.

Some companies have moved away from classic project funding, yet still tie product investment to annual cycles. That is a step forward, but it can still trap teams in an annual cycle of justification.

The issue is not simply whether the money is spent on a product. The issue is whether the organization sustains the same team's ownership of that product long enough for learning, support, improvement, and outcomes to compound.

If mindset is the invisible architecture of culture, funding is its visible proof.

The cadence encourages leaders to frame investment in terms of what can be delivered within 12 months rather than what the product needs over time.

The result is that even "product funding" can reinforce short-term thinking, especially when work is treated as a series of temporary initiatives rather than as a continuous line of ownership. Teams are pulled back toward features and scope rather than toward outcomes and stewardship.

Why it matters: This is how modernization and platform work stay protected from being labeled overhead. Stable product investment, sustained through durable teams, turns long-term system health into an intentional commitment rather than a discretionary expense that disappears under pressure.

How to Shift the Funding Model

Annual budgeting cycles often keep organizations tied to project-era thinking. McKinsey's research suggests that more mature product- and platform-based operating models are associated with stronger business outcomes and depend on more dynamic funding and reallocation.[13,14]

But how do you actually make that shift? You cannot simply tell the CFO to stop doing annual budgets. What leaders need first is a bridge: practical transition mechanisms that move the conversation away from fixed scope and toward capacity, outcomes, and value.

1. *Fund Products Through Stable Teams* Instead of approving every project as a temporary funding container, leaders begin by funding stable teams aligned to enduring products, platforms, or value streams.
2. *Review investments more frequently.* Rather than waiting for the next annual cycle, leaders create regular points to inspect progress, re-evaluate priorities, and adjust investment based on evidence.
3. *Measure value, not just output.* The conversation shifts from whether a predefined scope was delivered to whether the investment is producing useful outcomes, better flow, or stronger business results.
4. *Make funding decisions more reversible.* The goal is not to eliminate governance. It is to reduce the cost of being

wrong by making investment decisions easier to revisit as learning improves.

This is the point where I introduce a term I'll use later: the Value Stream P&L. Not a formal GAAP statement, but an internal leadership lens. It rolls up, by product or value stream, what you're investing (teams, platforms, cloud, support) and what you're getting back (revenue or outcome proxies). When you can see the margin by stream, portfolio decisions stop being political and start being economic.

Table 5.2—Project Budgeting vs. Product Funding

Dimension	Project Budgeting	Product Funding (Aspirational)
Cycle	Annual cycles, rigid approvals	Continuous investment in long-lived cross-functional teams
Investment Basis	Funds tied to deliverables	Funds tied to business outcomes
Justification	Business cases for fixed outputs	Investment cases for domains and learning

REALITY CHECK:
The Annual Budget Trap

Even organizations that adopt a product-centric approach often revert to annual budget cycles. Why? Because finance teams require predictability, boards expect yearly plans, and most companies lack the infrastructure to fund products continuously.

The breakthrough isn't eliminating annual reviews—it's changing *what* gets reviewed. Instead of justifying scope and deliverables, review outcomes achieved and outcomes targeted. The cycle stays annual, but the conversation shifts from "what will you build?" to "what impact will you create?"

At that stage, budgets still reset annually, but the conversation was beginning to move from features delivered to value realized. It was not a perfect model, but it was a meaningful step forward.

MEASUREMENT: FROM ACTIVITY TO IMPACT

As we established in Chapter 4, shifting from outputs to outcomes changes behavior. But metrics can't protect themselves. If teams are shuffled or funding resets annually, Flow Metrics and sentiment turn into noise.

Remember, numbers don't change cultures. Leaders do.

And structure determines whether metrics reinforce outcomes or collapse back into vanity.

Leadership Reflex vs. Discipline (Operating Model Edition)

Practice	Leadership Reflex	Leadership Discipline
Team Allocation	Redeploy teams or team members across products to chase capacity	"Keep teams intact, move work to them"
Funding	Annual resets tied to scope	Continuous investment tied to outcomes
Measurement	Activity and deadlines	Flow + outcomes + sentiment
Governance	Milestone reviews	Ongoing feedback and learning loops

Overcoming Resistance

Shifting operating models requires courage. Leaders must resist the comfort of project reflexes, such as redeploying teams, cutting developer experience investments, or reverting to annual project cycles. The hardest part is not technical, but cultural. It requires leaders to give up the illusion of control in favor of clear priorities, autonomy, and continuous learning.

Control feels safe until it strangles the very change we are trying to create.

Case Highlight — Vanguard's Value Stream Mapping Journey

Vanguard used a structured value stream mapping approach to break through operational silos and improve end-to-end processes. According to APQC, a research organization focused on benchmarking and process improvement, the effort brought leaders and employees together to visualize where customer value was created, where waste accumulated, and where handoffs slowed the work.[15]

APQC also notes that Vanguard's governance approach balanced business ownership with process team support and that the organization used a dedicated lead project coordinator for each value stream mapping initiative.[15]

What makes the example useful is not just the mapping itself, but the organizational learning it created.

APQC's summaries describe an environment marked by business silos, multiple handoffs, and significant quality problems before the work began.[16] They also report gains in collaboration across business groups and the elimination of a siloed work-distribution model.[16]

Viewed through the lens of this book, Vanguard's example shows how value stream thinking can strengthen visibility and coordination across the work, even in complex, legacy-rich environments.

Case in Point—Following a VSM Roadmap

One practical lesson from organizations adopting value stream management is that the work is more effective when it is sequenced. Rather than beginning with org-chart changes, leaders start by clarifying the vision, identifying value streams, mapping the current state, and building governance around the work.

The point is not to create disruption for its own sake, but to build momentum in a way the organization can absorb. Public VSM implementation guidance has framed this as a roadmap, moving from vision to identification, mapping, connection, and governance, rather than treating transformation as a one-time structural redesign.

That matters because moving from projects to value streams is not theoretical, but it rarely succeeds as a sudden reorganization.

It works better when leaders sequence the change, align stakeholders early, make the flow of work visible, and reinforce the model through governance. The roadmap helps organizations build momentum without creating unnecessary disruption.

PUTTING IT INTO PRACTICE

Shifting to a Product Operating Model isn't abstract. It requires deliberate, concrete actions. Here is how to begin:

If you are building a Product Operating Model:

- *Map one value stream end-to-end before reorganizing teams.* Understand where value flows and where it stalls. Do not re-org in the dark.
- Protect team stability long enough for trust and expertise to compound. Breaking teams apart resets knowledge and Flow.
- *Name the constraints you operate under.* Don't pretend regulations or budget cycles don't exist. Design for Flow *within* your realities, not someone else's.
- *Measure what matters from the start:* Flow, outcomes, and sentiment—not activity, utilization, or velocity.

If you are integrating two organizations:

- Document both operating models explicitly before deciding which to adopt. Make invisible structures visible.
- *Identify where architecture and team structure are misaligned.* Conway's Law will eventually surface these tensions; address them early.
- Give teams time to stabilize after integration before measuring Flow. Structural changes affect people first, outcomes second.

- Track sentiment alongside delivery metrics during the transition. The numbers alone won't tell you if the model is working.

In all cases:

- Start with one team, one value stream, one product. Prove the model works before scaling it.
- *Move work to teams, not people.* Protect the investment in long-lived, cross-functional structures. This doesn't mean locking people in place; movement for career growth or skills development is healthy. What erodes stability is shuffling people to chase capacity or project deadlines.
- *Resist the reflexes.* Don't redeploy teams to chase capacity, don't collapse product investment back into temporary scope, and don't revert to command-and-control under pressure.

KEY TAKEAWAYS

- *Structure determines whether practices stick.* Without the right operating model, Agile and DevOps collapse back into project reflexes.
- *Long-lived, cross-functional teams must be protected.* Break them apart, and you reset trust, expertise, and Flow.
- *Product funding remains aspirational.* Annual budgets are a structural anchor leaders must name honestly. The interim step is shifting the *content* of the budget review from scope to outcomes.
- *Metrics only matter if structure sustains them.* Without stable teams, Flow and sentiment become noise.
- *Leadership discipline sustains transformation.* Protecting teams, funding continuity, and resisting reflexes is the daily work of the leader.
- *Context matters.* Even in highly governed environments, leaders can adapt how controls are applied to protect Flow and outcomes.
- *Leadership mindset leaves fingerprints.* System architecture, team design, and role structures all reflect philosophy, whether distributed or consolidated.
- *The Price of Alignment is real.* Leaders must decide if a unified organizational chart is worth the cost of breaking architectural Flow.

- *Transformation is not about blame.* Different functions learn at different speeds. What matters is that organizations continue to evolve together.

Value Streams and Product Operating Models are not competing ideas—they are two halves of the same system. The product model defines *where* value is created; value stream management ensures that value keeps *moving.* One anchors outcomes, the other protects Flow. Together, they turn transformation from theory into traction.

But even the strongest structures depend on one thing: leadership. Operating models are only scaffolding; without discipline at the top, the weight of old reflexes pulls everything down.

This chapter touches on several disciplines that shape modern product delivery, but it does not go deep on each one. Readers who want to go further can use the appendices in two ways: Appendix C shows how we approached product and value stream identification in practice, while Appendix D, *Suggested Reading and Companion Practices for Modern Digital Product Delivery*, points to companion disciplines and suggested reading for deeper exploration.

The next chapter examines those reflexes, the ingrained habits that undo transformation even when the right structures exist. When leaders unconsciously revert to old patterns, such as reorganizing teams to chase capacity, cutting developer experience investments when budgets tighten, or defaulting to command-and-control when pressure rises, the structures collapse.

Chapter 6 presents a new perspective on leadership: not as vision and strategy, but as the daily disciplines that either sustain transformation or allow it to decay.

6
A New Way to Look at Leadership

"Don't move information to authority; move authority to the information." —David Marquet [1]

Leadership as a Hinge Point

The best structures in the world can't survive leadership that slips into old reflexes.

Chapter 5 showed how operating models provide the scaffolding, but leadership determines whether that scaffolding holds or buckles. Every transformation eventually hits this moment: when the system stops being a theory and starts depending on the people who run it.

The hardest part of becoming a senior leader was not learning new tools or practices; it was unlearning habits that had once made me successful. Experience matters, but without context and discipline, it can quietly undermine the very system a leader is trying to strengthen.

The difference between leaders who lift a transformation and those who derail it is not experience alone, but the discipline to apply it in context.

The best leaders use their background as context, not control. They draw on what they've learned to shape questions rather than dictate answers, adjusting their approach to the unique architecture, culture, and maturity of the organization they've inherited.

When I moved from Senior Director to VP, the focus shifted away from my own contributions and direct team oversight. The work shifted to structure and strategy: executing the vision, setting direction, and creating the conditions for teams to succeed. I was reminded often to "get out of the weeds," and I quickly learned that leadership at this level demanded constant learning, unlearning, and relearning.

It required humility and discipline to set the standard through my actions while relying on the suggestions and contributions of the teams I led. It also meant going toe-to-toe with legacy senior leaders, often in uncomfortable moments where I had to stand firmly behind the reasoning and the progress achieved so far, even when it would have been easier to back down.

The truth is, no matter how well-built your structures are, leadership behaviors ultimately determine whether they hold. Practices and metrics can be in place, but when leadership changes, reflexes return. The culture above you—your senior managers, peers, and executives you report to—can either accelerate progress or quietly undermine it.

I've seen both sides. At times, a single leader holding discipline created an island of modern practice that others could emulate. At other times, one reflexive decision unraveled months of work. This influence is why leadership sits at the hinge of transformation: it can either multiply the impact of structures or collapse them back into the very habits we thought we had left behind.

The Leadership Reflex vs. Leadership Discipline

Before we get to the tactics, we have to name the real obstacle: muscle memory shaped by past success.

These reflexes are the residue of habits that once produced results. In today's digital product economy, though, the same habits can become destructive. Leadership discipline is the daily decision to resist familiar reactions under pressure and choose behaviors that fit the market now.

The following patterns emerged repeatedly in practice. They are documented here as observable behaviors, not aspirational frameworks:

Domain	Reflex (Legacy Pattern)	Discipline (Modern Pattern)
Team Integrity	Redeploy team members or entire teams to chase capacity; treat them as interchangeable.	**Protect long-lived, domain-aligned teams.** Default to stable teams and move work into them. If cross-stream help is required, keep contributions bounded, time-boxed, and explicit. Quality and defect ownership always flows back to the original value stream.
Measurement	Ask for story points and deadlines to "prove" progress.	Hold the line on anticipated outcomes and Flow. Use sentiment as an early warning.
Governance	Milestone theater; approvals as control.	Frequent, lightweight reviews tied to outcomes and risk. Leaders remove friction, not add gates.
Funding	Annual resets push work back into projects (current reality).	Continuous, domain-based investment (explicitly aspirational for us).
Language	"Resources," "utilization," "velocity targets."	**People, outcomes, Flow.** Teams deliver, not individuals.

These friction points show up so consistently that they're worth naming.

The Five Transformation Friction Triggers

Most transformations do not fail because teams reject modern practices. They fail because leadership friction quietly overrides intent. These patterns rarely appear in strategy decks; they surface under pressure, when leaders revert to familiar reflexes and pull the organization back toward project-era behavior.

I have seen them repeat across companies and industries often enough that they are worth naming.

Domain	Reflex (Legacy Pattern)	Discipline (Modern Pattern)
Control Reflex (Authority Over Information)	Centralize decisions, add approvals, and tighten gates "until things stabilize."	Move authority to the information. Use guardrails to keep work safe, not gates to slow it down. Keep accountability close to the teams doing the work.
Funding Reset Gravity (Projects Rebranded as Products)	Annual resets force work back into scope, milestones, and short-term delivery commitments.	Invest in products through stable teams. Preserve continuity so ownership, learning, and outcomes compound rather than restart every cycle.

Metric Weaponization (Measurement Becomes Surveillance)	Turn metrics into scorecards for ranking, comparison, and punishment; teams optimize optics.	Treat metrics as steering signals and early warnings. Protect psychological safety so the system can tell the truth. Compare teams to their past selves, not to each other.
Decision Latency (Work Waits on Leadership)	Let tradeoffs stall, decisions drift, and priorities churn; teams wait and dependencies pile up.	Clarify decision rights and make tradeoffs explicit. Reduce decision latency by bringing decisions closer to the work and reviewing them frequently against outcomes.
Culture Discounting (Trust Treated as "Soft")	Defer trust, DevEx, and team health until "after we deliver," then wonder why performance is brittle.	Treat culture and DevEx as constraints on Flow. Invest early, protect feedback loops, and address engagement drops like production incidents.

Why it matters: These are not cultural preferences. They are operating disciplines that shape coordination cost, delay, and risk. When they hold discipline, Flow becomes predictable and trust compounds.

Your competitors face many of the same constraints. The advantage comes from how quickly and safely you can navigate them when guardrails are designed as reusable controls instead of one-off approvals.

LEADERSHIP PITFALL:
The Zombie Metric (When "Typing Volume" Becomes Talent)

It is tempting to think we left lines-of-code thinking behind years ago. We did not. I have seen versions of this metric resurface in high-stakes business reviews led by seasoned leaders.

In one case, a code-based analysis was used to rank 'top developers' and infer team value. The tool emphasized commit volume and churn history and produced a list.

The results were absurd.

The "top talent" list was dominated by engineers in a support-heavy function, where the daily work involved editing customer templates and managing text-heavy transformation changes. They produced massive volumes of change, but much of it was maintenance.

Meanwhile, architects and senior engineers doing the hardest work, designing the future platform, reducing coupling, mentoring teams, and solving scalability constraints, ranked near the bottom. Not because they lacked impact, but because they were spending more time thinking, designing, and guiding than typing.

Why it Matters: Legacy thinking doesn't disappear. It mutates and reappears when leaders feel exposed. When measurement confuses motion with progress, it penalizes the people building the future and rewards the work that keeps the past alive.

Every stalled transformation I've seen can be traced back to the same few leadership reflexes. Every successful one learned to recognize and counter them early.

When Senior Leadership Changes: A VP's Playbook

Leadership transitions above you are among the toughest stress tests for an operating model. You do not control them, but you can prepare for them.

Over time, I developed a repeatable playbook to help protect discipline when leadership changes. Each step acts as a guardrail against reflex.

1. Re-establish Intent in Plain Language

When new leaders arrive, the conversation can quickly slide back to status, velocity, and utilization unless the operating model is made explicit early.

I learned the hard way that principles alone were not enough. I needed a one-page brief that made the model visible: how teams were designed, what was non-negotiable, and how success would be judged. Without that, every debate drifted back into legacy language.

What follows is the operating brief I learned to use during leadership transitions. Leaders can adapt it to their own context.

LEADERSHIP ARTIFACT:
Re-Establishing Intent

Purpose

[Your organization's] investment in digital transformation encompasses not only technical aspects but also architectural and organizational considerations.

This document anchors the principles that protect that investment. It ensures that, as leadership or organizational structure changes, the organization continues to operate in ways that preserve autonomy, Flow, and architectural integrity.

Domain Model

Team design should reflect the microservice architecture. Small, cross-functional, self-managed, long-lived teams form the core of the model, each aligned to a domain, product, or value stream and owning its outcomes end-to-end.

- Because the platform is built on microservices, each team can own its service, domain, data, and deployment independently, enabling faster iteration, greater transparency, and continuous delivery of value.
- Teams are not interchangeable resources but are the engine of delivery, innovation, and learning.
- **Conway's Law Sidebar:** "Organizations design systems that mirror their communication structures."

— (Melvin Conway, 1968)[2]. The structure is not arbitrary; it mirrors our architecture. When organizational design shifts, architecture inevitably follows.

Non-Negotiables

- **Team Integrity:** Teams remain stable and domain-aligned. Work moves to the right team rather than moving people between teams.
- **Architecture Alignment:** Team boundaries and service boundaries evolve together. Structural changes must respect system design.
- **Accountability:** Every epic has an anticipated outcome that is reviewed against actual results. Accountability helps ensure the right things are being built and avoids the feature factory.
- **Decision Visibility:** Teams make most decisions collectively when priorities, risks, and trade-offs are visible to everyone.
- **Quality Is Non-Negotiable:** Defect backlogs are avoided by treating defects as unfinished work. If a defect is found in production, it is unfinished work from a previous iteration and must be addressed in the same iteration.

How We Measure Success

- **Flow:** How quickly and predictably work moves from idea to value.
- **Outcomes:** The business or customer results achieved against anticipated goals.
- **Sentiment:** The engagement, trust, and well-being of the teams delivering the work.

- **Architecture Health:** Degree of coupling, deployment independence, and time-to-restore.

Why it matters: This artifact is a friction reducer. It shortens onboarding time for new leaders, prevents decision thrash, and protects the system from reverting to milestone theater so teams keep delivering outcomes instead of rebuilding trust every quarter.

2. Clarify Decision Rights

Ambiguity is gasoline for reflex. New senior leaders often jumped straight into issues and asked the same question: who is accountable?

In a self-managed, domain-aligned model, accountability is distributed and must be made visible:

- *Domain team:* Accountable for implementation, quality, security, and the anticipated outcome.
- *Product leadership (the what):* Accountable for problem framing, scope, and the anticipated outcome.
- *Technical leadership (the how):* Accountable for system integrity, quality, and technical debt.
- *Scrum Master/Agile Delivery Manager:* Accountable for delivery Flow, team health, and working agreements.

In practice, I was ultimately accountable, but I had to make accountability visible in the moment by naming the decision owner, restating the non-negotiables, and anchoring the choice to the initiative's or epic's anticipated outcome.

Visible accountability creates trust. Without it, even good teams start guessing what leaders really want.

REALITY CHECK:
What Discipline Sounds Like

In the heat of a production issue or missed deadline, discipline shows up in language. Here is the difference:

The Scenario:
A key feature is at risk of missing a deadline.

The Reflexive Leader (Old Way):
"We need this by Friday. I don't care how you do it—add more people, work the weekend, get it done. Who is responsible for this delay?" (Signal: Anxiety, blame, input-focused.)

The Disciplined Leader (New Way):
"I see we are at risk for the Friday release. Let's look at the anticipated outcome. Is the scope we are trying to ship essential to that outcome, or can we trim it to hit the date? If not, do we move the date or accept the trade-off of quality risks? The decision belongs to the team, but I need transparency on the trade-offs." (Signal: Partnership, outcome-focused, guardrails.)

3. Preserve Delivery Leadership (Avoid the Hidden Debt)

It is tempting to fold facilitation and delivery stewardship into Product or Engineering Management. On paper, it looks efficient; in practice, it can erode psychological safety, increase cognitive load, blur incentives, and starve learning.

I have seen the difference clearly: when a dedicated Scrum Master or Agile Delivery Manager remained in place, retrospectives stayed honest, and Flow stayed visible. When the role was diluted, ceremonies collapsed into status updates and hidden debt accumulated.

Team Topologies warned us about cognitive load: if no one continuously tends to the Flow of work and ways of working, the system drifts and safety erodes.[3]

In Practice

During my tenure, we retained the Agile Leader role on many teams, though its empowerment and scope were not always standardized. I have also seen organizations absorb it into Engineering Management. That can work, but it raises the manager's cognitive load and weakens facilitation.

4. Unfreezing the Middle: From Bottleneck to Accelerator

The middle layer, especially the Directors and Engineering Managers, is the organization's transmission system. This is another layer that often determines whether transformation advances or stalls.

What gets called the 'frozen middle' is rarely a personnel problem. More often, it is a structural one: managers are told to empower teams while still being personally punished when dates slip. Under those opposing incentives, micromanagement becomes a predictable safety mechanism.

The role has to shift from driving the team to improving the system around the team.

This requires changing the questions senior leaders ask. When executives ask, "Is the team working hard?" or "Are they on track?", they trigger control reflexes. Discipline requires asking questions that uncover friction:

- "What systemic impediments are currently slowing your teams down?"
- "Do you have clear anticipated outcomes for the next epoch, or are you just managing tickets?"
- "How can we help you shield the team from the noise above?"

When the role shifts from taskmaster to flow-enabler, much of the resistance disappears. Many managers are eager to stop micromanaging; they just need organizational permission to lead differently.

Why it matters: The "frozen middle" is where strategy either turns into throughput or turns into delay. When you re-align incentives and questions at this layer, you reduce decision latency, protect focus, and increase the organization's speed-to-learning without adding headcount.

5. Reset Cadences

Old reflexes prefer big milestones and infrequent reviews. Discipline works through short cycles and continuous inspection.

When leadership changed, the teams quickly reset the rhythm by inviting new leaders to a Weekly Delivery Review focused on what had shipped to production and the outcomes it supported.

You can't preach new rhythms if you're still dancing to the old ones.

6. Expose Trade-Offs Early

When priorities collide, the reflex is to move people. Discipline protects team integrity first and moves the work into stable teams.

When cross-stream help is unavoidable, the trade-offs have to be made explicit:

- *Bounded Contribution:* The helping team takes on clearly defined epics, not open-ended support.
- *Accountability Clarity:* Once the work is in production, defects and quality ownership transfer back to the original value stream team.
- *Timebound and Visible:* Everyone acknowledges the cost of context loss and potential impact on Flow.

This approach reframes the conversation. Instead of debating which teams to raid, we debate scope, trade-offs, and outcomes.

7. Protect Language

Language reveals culture. If leaders start calling teams 'resources' and success 'utilization,' reflex has already entered the room. I learned to correct that language immediately.

'Resources' became 'teams.'

'Velocity' became 'Flow.'

'Features shipped' became 'outcomes realized.'

It felt petty at first, but it was not. Words pull people back toward the old system or forward into the new one.

8. Build a Two-Way Brief

New leaders arrive with their own methods to achieve goals. If those goals are not acknowledged, reflex takes over.

I learned to ask what outcomes they were accountable for that quarter and use that as a bridge to show how the domain-aligned, long-lived teams could help. When it worked, the posture shifted from defense to partnership.

This playbook is not a guarantee, but it does make discipline the default path.

Why it matters: Leadership practices are the operating system that determines whether strategy turns into outcomes or churn and rework. When leaders reduce decision latency and protect focus, margin improves, risk drops, and the organization learns faster with fewer "hero" moments.

These dynamics are not unique to one company. When leadership reflex overrides leadership discipline, the pattern repeats across industries from quietly to catastrophically. The following cases make that visible.

External Lessons: Failure Cases

We're not the first to wrestle with leadership reflexes. Several well-documented corporate failures illustrate how reflex over discipline can unravel even the most sophisticated structures.

- *Boeing 737 MAX: When schedule pressure outruns engineering judgment:* In its final report, the House Committee found that Boeing faced tremendous pressure to compete with Airbus, which drove efforts to cut costs, maintain the 737 MAX schedule, and avoid slowing production. The report also pointed to faulty technical assumptions, a lack of transparency by Boeing's management, and insufficient FAA oversight. The lesson is stark: formal structures and certifications cannot compensate for leadership behavior that allows schedule pressure and concealment to override safety signals.[4]
    - *The Flow Failure:* Viewed through a Flow lens, the breakdown was not only technical. It was also a failure of information flow. The committee documented a culture of concealment in which crucial safety information did not move at the speed or with the visibility it required. In a healthy product operating model, safety data functions as a stop-the-line signal. In this case,

it was muted by pressure, opacity, and poor escalation.

- *Wells Fargo: When output targets poison culture:* For years, employees operated under aggressive "cross-sell" goals. Measurement is not neutral; when leaders glorify activity over outcomes, culture degrades, and customers pay the price.[5]
    - *The Metric Failure:* This is the ultimate "Goodhart's Law" failure: "When a measure becomes a target, it ceases to be a good measure." Leadership focused entirely on a proxy metric (account openings) rather than the value stream outcome (customer financial health). It proves that without the Sentiment and Outcomes layers of our measurement model (Chapter 4), simple numeric targets invite gaming.
- *Spotify: When branding outruns reality:* The "squads and tribes" story traveled faster than its reality. Internally, leadership reflexes around centralization and heroics often undercut the famous autonomy model. The caution is evergreen: naming an operating model isn't the same as leading with Discipline.[6]

SUCCESS STORIES: DISCIPLINE AT WORK

In successful transformations, the turning point is rarely the framework alone. The shift takes hold when senior leaders reinforce new ways of working through their own behavior, such as setting clear goals, protecting team autonomy, and modeling the discipline they expect from others.

Looking at organizations like GlobalTech Inc., the turning point wasn't the choice of framework but the behavior of senior leaders. When leaders modeled autonomy-first behaviors, teams were empowered to take ownership of their outcomes.[7]

Similarly, John Deere's Global IT group scaled Scrum and Scrum@Scale across its 500 teams through a leadership-backed transformation with clear enterprise goals. The results included a 165 percent increase in output, while eNPS rose from a 42-point baseline to 65.[8]

These stories reaffirm the central point: reflexes may be seductive, but discipline reinforced by leadership produces lasting results. I did not learn that only by watching other organizations. I had to confront the same pattern in myself.

My Own Reflex: From Pagers to Leadership

In my early years at DHL, I carried a pager. When it buzzed, I reacted immediately, diving into logs, suggesting fixes, and pulling in whoever I thought might solve the problem fastest.

That urgency was necessary at the time. Downtime meant disrupted global logistics. Speed was survival, and that reflex was rewarded. I still remember the mental weight of that pager.

As I grew into leadership, those habits followed me. Early at Parchment, issues weren't paged to engineers; they were escalated directly to senior managers like me. My instinct was still to react quickly: propose solutions, call in top engineers, drive toward immediate fixes.

But the context had changed. By jumping in too fast, I robbed teams of the space to diagnose root causes, weigh options, and build confidence in their own judgment. I came to see that the reflexes that once made me successful were now undermining me.

I was no longer at my best by being the first responder. My job had become creating the conditions for others to lead the response well. Real discipline meant slowing down, asking better questions, and leaving room for others to solve the problem. Reflex has built my career. Discipline would sustain it.

Strategic Partners Have a Backbone, Not a Wishbone

Moving from a reflexive cost center to a disciplined strategic partner requires a fundamental shift in posture. You cannot be a partner if you are afraid to confront reality.

I often tell my teams that I have no interest in transformation theater: organizations that claim they want to transform, yet repeatedly choose comfort over the discipline real change requires. The difference usually comes down to anatomy:

- *The Wishbone Leader* hopes for alignment, accepts impossible deadlines to avoid conflict, and shields the business from the reality of engineering constraints. They treat improvement as a "nice to have."
- *The Backbone Leader* drives a culture of respect by *confronting reality early*. They engage in healthy debate, keep an open mind, demand shared accountability, and refuse to hide the work.

I want a leadership team with the confidence to take smart risks and the humility to learn fast. That requires a backbone: improvement is the job, not a side project.

What Leaders Actually Do (When It Works)

"Execution is intelligence. It isn't theory—it's the operating rhythm where decisions turn into outcomes."

In practice, this is what disciplined leadership looks like. It rests on four enduring pillars:

1. *Enterprise-level Software Quality and Resilience:* Protecting Flow and Quality Metrics. This means protecting architecture runway, the foundational technical work that keeps future delivery possible, while also reducing toil and maintaining feedback loops.
2. *People Engagement:* Safeguarding psychological safety, autonomy, and purpose. Engagement dips are treated like production incidents.
3. *Retention and Development:* Leaders sponsor stretch work and ensure that support is available. Promotions are tied to demonstrated impact and behaviors, not heroics.
4. *Skills and Capabilities:* Creating space for experimentation. Disciplined leaders ensure teams don't just deliver today's features but build tomorrow's capacity through learning budgets and shared libraries.

When every engineering leader shares responsibility for improving the system, productivity stops being an initiative and becomes an operating principle. That is what keeps DevEx from becoming someone else's problem.

When leaders focus here, they multiply the system's capacity to adapt and thrive. The opposite is equally true: neglect these pillars, and the system withers.

Collaboration as a Force Multiplier

Disciplined leaders do not try to transform alone.

> **LEADERSHIP PITFALL:**
> The Isolated Change Agent (The Solo Crusader Trap)
>
> After a talk where I spoke about transformation friction, I received a message from a leader at a large company.
>
> "Phil, I'm drowning," he wrote. "Upper management refuses to discuss new ideas. The status quo is crushing my team. Everyone is afraid to speak up. How did you handle it? Did you bring in Agile Coaches to help fight these battles?"
>
> He was trying to solve a political problem with a tactical response, hoping process expertise would overcome executive resistance.
>
> I told him the truth. Resistance was never clean or comfortable. And an Agile Coach cannot save you from leaders who do not want to change. This is the Solo Crusader trap. Leaders burn out trying to win with logic and process charts while their opponents are fighting with politics and inertia.

My advice was simpler: stop trying to convert everyone. Find the one or two enlightened stakeholders who actually want the outcome you are promising. Build executive sponsorship. Then move forward with discipline, not personal heroics.

Why it matters: You cannot transform an organization from an island. Without executive sponsorship, the change remains personal effort, not organizational transformation. And when that happens, the change agent does not lead transformation. They absorb the cost of it.

Some of the most significant turning points in my career didn't come from frameworks; they came from honest conversations in trusted spaces.

When leaders from non-competing organizations let down their guard and share what *isn't* working, they find the courage to challenge their own defaults.

You must build this same space for your teams. When you create an environment where it is safe to say "I don't know" or "This isn't working," you replace the fragility of heroics with the resilience of collective intelligence.

Key Takeaways

- *Leadership behavior is the load-bearing wall.* Structures and metrics crack without leaders who protect long-lived teams, anticipated outcomes, and visible Flow.
- *Reflex vs. discipline is a daily choice.* Under pressure, protect team integrity first. Default to moving work into stable teams rather than moving people.
- *Delivery leadership is not overhead.* A dedicated Scrum Master or Agile Delivery Manager reduces cognitive load and maintains system balance. Cutting it creates hidden debt that manifests later as delays and defects.
- *Culture is measurable and material.* Psychological safety and generative cultures correlate with delivery performance. Treat engagement drops like incidents.[9]
- *Lead by system, not heroics.* Distinguish current practice from the target state to keep the pressure on growing discipline over time.

Leadership choices do not exist in a vacuum. They shape structure, incentives, and behavior. Those forces then determine whether alignment holds or begins to fracture.

Chapter 7 zooms out to strategic alignment: how to connect technology, product, and business so outcomes survive the organizational gravity that pulls everyone back to projects and outputs.

7
Overcoming Barriers to Strategic Alignment

"Alignment is a key ingredient to elite performance because without it the best strategy in the world cannot be executed." —Urban Meyer [1]

When Leadership Demands the Wrong Insights

The strongest systems begin to fracture when leadership and strategy drift apart. The best intentions of the C-suite mean little if goals, metrics, structures, and decisions remain aligned with outdated assumptions.

That is where transformation breaks down, not in the code or the culture, but in the space between intent and execution.

One of the most damaging leadership failures I have seen in digital transformation is the failure to connect Agile, DevOps, and Lean practices to outcomes that business leaders already understand. I described one version of this earlier in Chapter 1.

Here, the point is what that moment revealed about strategic alignment.

In a high-stakes executive conversation, a question surfaced that sounded simple but exposed a deeper gap: "How can we measure developer productivity? Can we track engineers the way we measure sales reps or support agents?"

This is a legitimate question, and I have encountered it across multiple organizations. Senior leaders want legibility.

They want a metric that can be rolled up, compared, and managed, something that behaves like a quota in a world where knowledge work resists simple quantification.

That instinct is understandable, but it becomes dangerous the moment the metric stops being a signal and becomes the goal.

Once leadership asks for a 10 or 15 percent increase in productivity, the system begins to bend around the number rather than around the outcome. Teams start optimizing for what is easiest to count and easiest to display, so visible output may rise even when meaningful progress does not.

That is where alignment quietly breaks. The business starts asking for a number it can manage, while teams adapt their behavior to meet that number.

More tickets can mean smaller slices of the same work. More commits can mean more activity without more value. More output can mean more code when simplification would have served the customer better. The dashboard looks healthier, but the system may be drifting further from the outcome it was supposed to serve.

That moment exposed the alignment gap between what leadership wanted to measure and what the system actually needed to improve.

Leadership was reaching for output proxies, while I was trying to lead with system truth: Flow, bottlenecks, sustainability, and whether those investments were producing outcomes the business could actually stand behind.

This is why I now advocate not only for Flow, how efficiently work moves through the system, but also for Realization: whether an improvement or investment actually achieved its anticipated business or customer outcome.

The Persistent Struggle for Alignment

The struggle for alignment has shadowed software engineering since the beginning. In the finance world, engineering is often viewed as a cost center on paper. That is how accounting works: salaries, tools, and infrastructure are logged as expenses, much like HR or office rent.

But labels on a ledger don't tell the whole story.

In digital product companies, engineering is not just an internal service. It creates the product, delivers customer value, and drives profitability. It is inseparable from the business itself. Whether finance calls it a cost center or not, the reality is that engineering determines how competitive the company is, how quickly it can respond to the market, and how much profit it can sustain.

The problem is that many leaders still operate through the old lens. If they view engineering as overhead, they optimize for efficiency per developer, squeezing costs rather than growing value. If they see engineering as a strategic partner, they optimize for flow and outcomes, investing in structures that accelerate business results. Strategic alignment depends on which story leadership chooses to believe, because that story shapes the incentives, measures, and structures the organization builds around engineering.

Why Alignment Still Fails

Alignment erodes when goals, incentives, structures, and decision-making drift out of sync, even in organizations full of committed people.

- Legacy Mindset: Technology is still seen as a cost center. Leaders ask how much it costs per developer, not how much value it creates per team.
- Siloed Structures: Product, engineering, and operations still work in functional tracks. Handoffs persist. Teams are treated as interchangeable capacity rather than outcome owners.
- Misaligned Incentives: Organizations reward predictability and output rather than outcomes. Product managers are judged by the features shipped. Engineers by individual velocity. Executives by quarterly predictability. None of these, by themselves, drives impact.
- Fragmented Learning: Even when teams discover better ways of working, organizations often lack a mechanism to spread those lessons. Learning stays local, and declared priorities rarely become operating habits.

I call this Alignment Theater: priorities are declared, but incentives, measures, and structures do not actually change.

Why it matters: Misalignment across goals, incentives, structures, and execution is not just a messaging problem. It risks investing in the wrong work, increases coordination overhead, and pushes leadership into late-stage trade-offs when the only lever left is cost-cutting.

Strategy Theater

I have walked out of off-sites with a strategy deck, notes, and a room full of excitement. Six weeks later, nothing had shifted. Teams still reported velocity, executives still tracked utilization, and product managers were still judged on roadmap distance. The strategy sounded inspiring, but because incentives hadn't changed, behavior hadn't either.

Tool-First Thinking

At one point, the organization invested heavily in building an internal Flow metrics dashboard. It looked impressive, but it was fragile. Keeping it alive required three to four people's salaries and effort, far too much overhead for something outside the organization's core expertise.

The dashboard made bottlenecks visible. It showed throughput, cycle time, and wait states, and that visibility mattered. Teams could finally see where the system was slowing them down.

But it only told part of the story. It did not show what happened after release. Did the investment reduce support costs, improve retention, or change customer behavior?

Without visibility into Realization—the actual business or customer impact after release—the story remained incomplete. The dashboard showed efficiency, but not return on investment.

That is the difference between a dashboard that informs teams and one that earns executive confidence. Flow shows delivery efficiency. Realization shows whether the investment delivered the result it was intended to produce. Without both, organizations eventually optimize for optics instead of outcomes.

THE HIGH COST OF MISALIGNMENT

Nordstrom's public DevOps case, as described by Courtney Kissler, illustrates a hard truth about transformation inside established organizations.[2]

The technology organization had been optimized for cost, annual planning, and large-batch releases, and by traditional measures it was performing well against schedule, budget, and scope. But those strengths were no longer enough once the business needed greater speed and a faster response to customer needs.

Nordstrom's response was to test a different operating model in targeted areas. In its mobile application work, the company shifted to a dedicated product team, continuous planning, and a single prioritized backlog tied more directly to customer needs. Public accounts reported that over the following year, the team doubled the number of features delivered per month while cutting defects in half.[3]

By any reasonable standard, that was evidence that the new model was working. But one of the painful realities of transformation is that success in a few teams does not guarantee that the broader system will protect or sustain it.

When leadership, funding, and organizational design remain tied to the older model, the new way of working can be treated as an exception rather than the future. That is what makes the Nordstrom story so instructive here: even credible progress can be marginalized or reversed when alignment around the operating model never fully takes hold.

Alignment is shaped not only by goals and structure, but also by how organizations learn under pressure. Strategy only holds when people can surface reality, learn quickly, and respond without fear.

Etsy's public engineering writing offers a different but related lesson. In Code as Craft, John Allspaw described Etsy's approach to learning from accidents and mistakes. In Etsy's 2016 debriefing guide, he pointed back to that earlier 2012 work while emphasizing that postmortem debriefings should be treated first as learning opportunities, not as blame sessions or simplistic fixing exercises. The guide argues that organizations create only an illusion of understanding when they rush to a single root cause instead of gathering multiple perspectives and placing objective data in context.[6]

My takeaway is that alignment is affected not only by goals and metrics, but also by how an organization interprets failure. When leaders treat incidents as opportunities for broader learning, they create better conditions for trust, resilience, and continuous improvement.

This is what makes alignment more than a structural exercise: strategy only holds when people can surface reality, learn quickly, and respond without fear.

WHEN STRATEGY IGNORES REALITY

Chapter 5 explored the structural fallout of this: the collision between a distributed microservices architecture and a consolidated, manager-led organizational chart. In practical terms, teams were combined to fit a 'span of control' model, breaking the carefully built domain boundaries.

But the *lesson in alignment* here is more profound than the org chart. The failure wasn't just about reporting lines; it was a failure of *strategic translation.*

In my experience, mandates to consolidate teams are usually driven by a desire for organizational symmetry and visible efficiency. Leaders look at the spreadsheet and see an opportunity to standardize spans of control or normalize the ratio of engineers to managers. What can get lost in that view is that some small teams are not inefficiencies at all. They are the structural glue holding a distributed system together.

This is a language gap as much as a strategy gap. One side speaks the language of Flow, speed, autonomy, and domain fidelity, while the other speaks the language of managerial control, ratios, standardization, and predictability.

This brings us to a painful truth about strategic alignment: *Experience can sometimes be the enemy of context.*

WHEN EXPERIENCE REPLACES CURIOSITY

Across many organizations, I've observed a recurring dynamic when experienced leaders step into unfamiliar systems. Teams try to explain the journey: how they re-architected the system, separated domains, built independently deployable services, and aligned teams to that architecture.

And sometimes, the response is a version of: "My playbook works. I don't need to understand this."

It is a pattern that Liz Wiseman describes in Multipliers: leaders who rely on their own expertise can inadvertently shut down the intelligence around them.[7] It isn't malice. It is the comfort of a proven playbook applied to an unfamiliar context.

When leaders have built their careers in monolithic environments, with centralized ownership and large teams, it is natural to trust that playbook. But playbooks are contextual, not universal. Modern systems are distributed. They depend on clearly defined domains and subdomains, independent services with stable contracts, and localized decisions made close to the work.

Those conditions require structures that look different from traditional span-of-control models. Experience is valuable, but experience without curiosity becomes a constraint—especially in distributed systems.

This led me to believe that *architectural curiosity* is now a core leadership competency. Leaders don't need to design microservices or draw sequence diagrams. But they do need to understand the implications of architecture on how teams are structured, how work flows, how value is delivered, and what will break if they force uniformity on top of it.

Practices That Create Alignment

Breaking the cycle of 'Alignment Theater' requires replacing project reflexes with product discipline.

1. Shared Objectives (OKRs & EBM)

A common measurement framework does not give leaders all the answers. It helps goals, measures, and decisions stay connected.

Two practices are especially useful here: Objectives and Key Results (OKRs) and Evidence-Based Management (EBM). Together, they create a shared language for discussing progress, value, and outcomes across leadership, engineering, and cross-functional partners.

But shared language alone does not solve the translation problem between strategy and execution.

One of the recurring struggles I saw was that Product Owners and teams were often asked to identify the primary anticipated outcome of a piece of work only after it had already reached them, even though that outcome should have been made clearer upstream.

I came to understand this problem more clearly later through Klaus Leopold's *Flight Levels*, which describes three connected levels of work: strategy, coordination, and operations.[8]

I found them useful because they gave language to a gap I had already experienced. In many organizations, strategy exists at the top, and teams work hard at the ground level, but the coordination layer in the middle is either too weak or too implicit. That middle layer is where priorities should be translated, dependencies surfaced, tradeoffs made visible, and cross-functional work aligned before confusion reaches the teams.

When that coordination layer is weak, teams are left trying to define purpose too late. They inherit work items, roadmap commitments, or delivery requests without a clear throughline back to the business result the work is supposed to influence.

In healthier systems, teams do not invent outcomes in isolation.

A division-level key result can serve as the basis for an operational team's objective, which can then shape the anticipated outcome for the initiative, epic, or change being delivered.

When that chain is clear, Product Owners and teams can explain why the work matters and what result it is meant to create. When it is not, they fall back to output language, activity, or vague delivery goals.

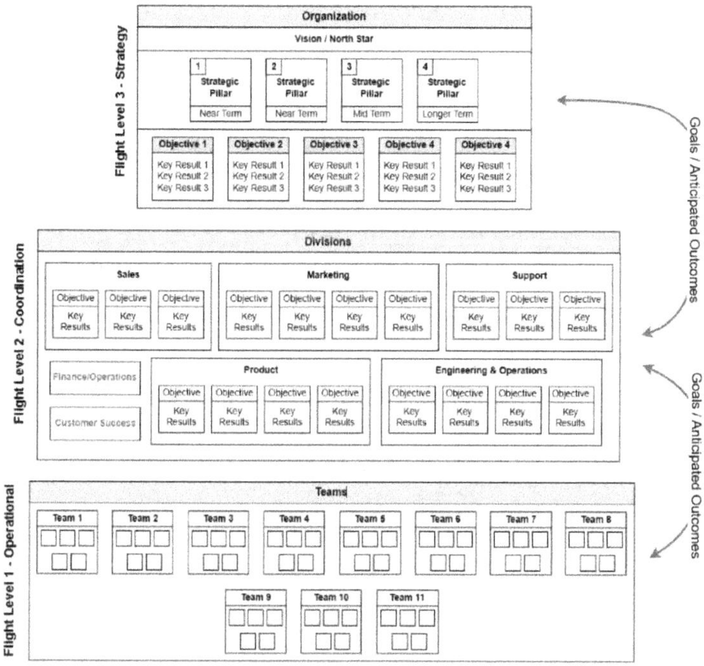

Figure 7.1. *Connecting strategy to team outcomes through the middle layer.* Author illustration based on the *Flight Levels* model described by Klaus Leopold and Siegfried Kaltenecker. This illustration shows how strategic objectives and key results can cascade through divisions into team-level objectives. In this model, a division-level key result can become the basis for an operational team's objective, helping product owners and teams define the anticipated outcome of the work more clearly.

In my experience, executive-level OKRs helped connect leadership priorities to five company-level goals. I created department-level OKRs that translated those goals into team-level priorities and measures. But because of continued organizational change and roadmap pressure, they did not take hold across all teams during my tenure.

What did stick was the discipline around anticipated outcomes. By that point, every initiative or epic in Jira required a defined anticipated outcome tied to a key result. That requirement forced teams to articulate why the work mattered before it started.

2. Unified Metrics: Flow and Realization

Flow and Realization tell one story together. Flow shows system health and delivery movement. Realization shows whether the work achieved its intended business or customer result.

Flow without Realization measures motion. Realization without Flow misses the system conditions that produced it. Together, they align.

3. Purpose-Driven Development

Every initiative or epic should carry an anticipated outcome. The more important question is not simply what will be delivered, but what will change.

- Feature: "Add new transcript upload flow."
- Anticipated Outcome: "Reduce average support tickets by 25 percent in the next quarter."

This practice turns backlog items into hypotheses. It forces leaders to declare intent in business terms before work begins.

- The anticipated outcome requirement the teams documented in *Jira*, a popular project management and issue-tracking software, became one of the most effective alignment tools. Product managers initially resisted it as overhead, seeing it as extra work that slowed down backlog grooming. Engineering leaders and Agile Delivery Managers championed it because it helped teams understand why their work mattered.

4. Closing the Loop with Realization

One of the hardest parts of keeping goals, delivery, and business results connected is proving impact after release. Did the change reduce support tickets? Increase renewals? Deliver the business result that had been expected?

At that stage, this had not yet become a discipline across all teams. Documenting actual outcomes remained uneven. But the pressure was intentional. By asking for actuals, I kept the expectation visible: it is not enough to ship work; teams have to learn whether it created the value predicted.

Even engineering-first organizations with mature developer experience and strong executive sponsorship struggle to close that loop. That is exactly why Realization cannot be optional. It is the discipline of proving whether the shipped work created the value it was expected to create.

The Rhythm of Respect: A Communication Checklist

A culture of respect is built less through slogans and more through a steady communication rhythm.

Information hoarded erodes trust. Information shared accelerates execution because people aren't wasting cycles guessing where they stand. On a regular cadence, I recommend covering five points with senior stakeholders and partners:

1. *The Environment:* What does the broader market look like right now?
2. *The Decisions:* What strategic choices have been made?
3. *The Changes:* What has changed since the last conversation?
4. *The Reality:* What is going well, and honestly, what is not?
5. *The Impact:* What does this mean for the teams and the work right in front of them?

You also need a mechanism for employees to ask questions that you must actually respond to. When you treat people like professionals who can handle the truth about trade-offs, you build an environment where they can move fast without constantly looking over their shoulder.

VALUE STREAM THINKING: THE ALIGNMENT LENS

Value Stream Management (VSM) provides the scaffolding for this discipline. This shift from feature-based to flow-based delivery, as established in *Accelerate* and reinforced in *Project to Product*, fundamentally changes how organizations measure success.[4,5]

When you use Value Stream thinking, you stop measuring "Departments" (Dev, QA, Ops) and start measuring "Flow."

- *Traditional Model:* Features delivered = Success. Velocity is the metric.
- *Value Stream Model:* Success is measured through both Flow and Realization.

Value stream mapping exposes the gaps: stalled ideas, wasteful handoffs, and funding cuts that sever outcomes. Flow Metrics provide the system lens. Realization connects those delivery signals back to business results.

Agile Delivery Managers as Flow Advisors

Over time, I evolved the facilitation role from traditional Scrum Master into what I called an Agile Delivery Manager, with a deliberate focus on value streams and flow. Their responsibilities shifted from primarily facilitating ceremonies to interpreting flow data, surfacing risks, and connecting delivery work to business and customer outcomes.

Instead of reporting activity, Agile Delivery Managers ensured leadership conversations focused on outcomes. They became translators between the system-level view of flow and the business-level view of results. It was a visible step toward treating engineering as a strategic partner rather than a cost center.

Why it matters: At scale, alignment becomes credible only when leaders can show how changes in flow and delivery behavior translate into measurable business impact.

CASE STORIES: ALIGNMENT AT SCALE

ING: Shifting from Projects to Value Streams

Beginning in 2015, ING reorganized roughly 3,500 employees at group headquarters into cross-functional squads and tribes, inspired in part by practices observed at technology firms such as Spotify. The intent went beyond adopting Agile ceremonies. It was a shift toward end-to-end, multidisciplinary teams organized around customer journeys and faster delivery.

McKinsey's analysis highlighted three core changes:[9]

- Cross-functional squads brought together product, commercial, UX, data, and IT capabilities into end-to-end delivery units.
- Tribes, quarterly business reviews, and new coordination mechanisms created coherence across domains.
- Roles, governance, and the people model were reworked to reduce handoffs, weaken siloed project structures, and emphasize customer impact.

The lesson is clear: this kind of alignment did not happen by accident. ING changed team design, governance, and management practices together, creating conditions for business and technology to work from a more unified model.

John Deere Global IT: Moving Beyond Output

John Deere is often cited as a major enterprise-scale Agile transformation. Public case materials describe a multi-year effort spanning roughly 500 global technology teams, with reported gains including a 165% increase in output and a 63% reduction in time-to-market.[10]

But the story did not stop with throughput. John Deere leaders framed the transformation around "speed to outcomes," not speed for its own sake, and later public summaries of their ongoing journey described an evolution from outputs to outcomes to impact.[11]

Other John Deere case material reinforces that shift. TBM Council's write-up describes moves from annual portfolio planning to continuous portfolio governance, from IT initiatives to digital products, and toward value-focused product cost reporting and investment conversations.[12]

The broader lesson is that alignment at scale becomes credible when leaders can connect delivery improvements to business value, investment choices, and customer outcomes over time.

Reflection: Alignment Is Never "Done"

Alignment is not a milestone you reach; it's a state you maintain. It is a discipline leaders must renew constantly. Structures drift. Incentives revert. Silos reappear.

Leaders who assume shared goals and incentives will hold after a workshop are the same ones who later ask why strategy isn't reflected in delivery. The real work is relentless: keeping incentives honest, making outcomes visible, and holding the line when short-term pressures tempt leaders back into project reflexes.

It is about proving that software engineering serves as a strategic partner in the business.

When engineering is treated as a cost center, organizations count productivity units and optimize for efficiency per developer. When engineering is treated as a strategic partner, they design teams around domain fidelity, measure outcomes over outputs, and link every initiative to customer and business impact.

In today's environment, where speed and adaptability define survival, that choice is existential.

Key Takeaways

- *Incentives Drive Behavior:* If you reward outputs, you will get activity. If you reward outcomes, you will get impact. Alignment fails when strategy says one thing but incentives measure another.
- *Flow + Realization:* Measuring Flow is not enough. You must also measure whether shipped work achieved its anticipated outcome.
- *Conway's Law is a Constraint, Not a Theory:* Organizational structures must mirror architecture. When you force a distributed system into a monolithic "span of control" leadership model, you break the system's ability to flow.
- *Architectural Curiosity:* Leaders do not need to code, but they must understand how their org chart impacts their architecture. Experience without curiosity is a liability in modern systems.
- *Purpose-Driven Development:* Every Epic requires an "Anticipated Outcome." This simple rule forces alignment before code is written.

As hard as strategic alignment is, it is only part of the story. Structures, dashboards, and incentives matter, but they are not enough on their own. Alignment only works if the people inside those structures have the trust, energy, and safety to sustain it.

Over the years of leading teams through transformation, I have learned that alignment cannot be assumed. Even when metrics look strong or structures appear solid, it frays quickly if teams are exhausted by deadlines, unsettled by constant change, left without support, or disconnected from purpose.

That is why culture is the multiplier. The next chapter turns to the people, the ultimate engine of sustainable performance, and how to build a culture that survives the pressure of transformation.

8
INVESTING IN PEOPLE & CULTURE FOR HIGH-PERFORMING TEAMS

"When people are financially invested, they want a return. When people are emotionally invested, they want to contribute." —Simon Sinek[1]

One of the deepest lessons of leadership is that people are not peripheral to change; they are at its center. Strategy may set direction, and technology may enable scale, but people determine whether either one becomes real. In my experience, the harder part was rarely the technology itself. It was building the conditions in which people could do their best work together.

SHIELDING AS A CULTURAL ACT

One of the most defining lessons I learned about culture didn't come from a workshop or a manifesto. It came from watching leaders act as shields.

Early in my career, I worked in an environment where the pressure was constant. Executive demands were volatile, deadlines were often divorced from reality, and conflicting strategic pivots arrived weekly. By all logic, the engineering teams should have been chaotic, stressed, and fractured.

But they weren't. They were focused. They were calm. They delivered.

The difference was the layer of leadership above them. These leaders didn't stop the pressure—they absorbed it. They took the heat from the boardroom, reframed the chaos into context, and passed down only what the teams could act on. It wasn't about hiding reality; they were transparent about the business challenges we faced. But they filtered out the noise: the panic, the politics, and the conflicting directives that paralyze execution.

That protection *is* culture.

It told the teams, "We trust you to do the work, and we've got your back on the rest."

This shielding worked because it wasn't a trick; it was backed by a heavy investment in core principles. Even when new senior leadership arrived and pressure from above threatened to shift priorities, those foundations remained intact.

The result was a retention rate that consistently exceeded industry benchmarks for software engineering teams, sustained over nearly a decade. When people did leave, it was rarely out of frustration. They left because they had outgrown us, moving on to opportunities we genuinely couldn't match at our scale.

Exit interviews told a consistent story. People spoke with pride about their contributions. They loved their teams. They loved the work. To us, that feedback validated a core hypothesis: investing in people, through career development, upskilling, and genuine autonomy,[2] wasn't just an HR retention strategy. It was a business continuity strategy.

At that stage of growth, salaries were competitive, but still well below the top of the market. Yet, average tenure for software developers significantly exceeded industry norms. These weren't people hanging on due to complacency; they were some of the strongest performers in the organization. They were the engineers who knew the legacy monolith *and* the modern microservices architecture. They were the ones recruiters aggressively targeted.

They stayed because the culture allowed them to do their best work.

Several times, high performers left for a significant pay bump at another organization, only to return a year later. They told us that the "tax" of working in a dysfunctional culture—the bureaucracy, the lack of safety, the inability to ship—wasn't worth the raise.

This cultural retention had a hard financial impact. These long-tenured teams formed the backbone of the organization as revenue grew more than twentyfold during my tenure, driven by a combination of organic growth and strategic mergers.

Their stability enabled us to navigate that complexity, resulting in a transaction publicly valued at approximately $835 million.[3]

To be clear: Engineering did not create that value alone. Growth at that level also requires strong sales execution, product strategy, and operational discipline. What made that growth scalable was a high-performing platform rather than a legacy system that dragged the business down.

In every evaluation, the review went beyond financials to include software delivery processes, practices, and governance. Passing that technical diligence proved that our system was an asset, not a liability of technical debt.

Culture didn't just sustain our morale; it sustained the engineering discipline that underpinned our valuation.

WHO 'PEOPLE FIRST' REALLY INCLUDES

I am people-first, but not in the narrow sense the phrase sometimes suggests. When I say "people," I do not mean only the engineers on payroll. People include both employees and customers.

You cannot sustain a great customer experience built on the backs of a burnt-out engineering team. Eventually, the internal friction leaks out to the customer. Conversely, you cannot have a great employee culture if the product is failing the customer.

Sustainable delivery only works when *both* experiences improve simultaneously. If you ignore one, you inevitably break the other.

THE MULTIPLIER EFFECT

The previous chapters built the operating foundation: structure, leadership discipline, and strategic alignment. But none of that works if people are afraid to tell the truth."

But here is the hard truth: You cannot measure Realization if your people are afraid to tell you the truth.

If a team fears punishment for a failed experiment, they will not report that a feature failed to achieve its outcome. They will report "Done." They will game the metrics to show "Green." You will have a dashboard that says success, and a bank account that says failure.

Culture is the operating environment. It is the road surface. You can have the right structure and capable leaders, but fear still turns the road to ice.

Gallup's Q12 meta-analysis found that business units in the top quartile of employee engagement showed 23% greater profitability and higher productivity than those in the bottom quartile.[4] Google's Project Aristotle identified psychological safety as the most important dynamic of effective teams.[5]

Organizations can redesign operating models, adopt a value stream metrics platform for flow metrics, and align in centives, but without safety, engagement, and trust, even the best systems revert to output-chasing and attrition.

BEYOND "NICE": THE WESTRUM CONTINUUM

For a long time, I struggled to define 'good' culture precisely. Ron Westrum's work helped me see it as a pattern of information flow.

Westrum defined three types of organizational cultures: *Pathological*, *Bureaucratic*, and *Generative*.[6]

My role as a leader was effectively to move teams from the left side of this model to the right.

1. *Pathological (Power-Oriented):* Information is hidden. Messengers are shot. Responsibilities are shirked. Collaboration is discouraged.
2. *Bureaucratic (Rule-Oriented):* Information is ignored. Messengers are tolerated but not heard. Responsibility is compartmentalized ("That's not my job").
3. *Generative (Performance-Oriented):* Information is actively sought. Messengers are trained. Risks are shared. Collaboration is encouraged.

I have seen the *Pathological* state firsthand in mergers and acquisitions (M&A) scenarios. I've seen pathological cultures where engineers concealed quality issues until the day of release because raising concerns was treated as failure rather than discipline. The result was never a faster launch. It was a launch that exploded in production, requiring weeks of remediation. The "speed" was an illusion created by fear.

The *Bureaucratic* state is often where innovation goes to die. This is where the answer is "No" simply because "we don't have a policy for that."

The division navigated this tension when *GitHub Copilot* and *ChatGPT* exploded onto the scene in early 2023. The pressure to adopt them immediately was high, but the risk of IP leakage into public models was real. A *Bureaucratic* culture would have issued a blanket ban and walked away, choosing safety through stagnation.

Instead, the response reflected a generative culture. The idea was paused, not abandoned, to protect code, personally identifiable information (PII), and intellectual property (IP). That time was used to educate the teams and establish the necessary guardrails. In a generative culture, a pause creates the conditions for safer, faster progress.

My goal was to move teams toward that generative state, because that is where Realization becomes possible. In a generative culture, when a deployment fails, the team doesn't ask "Who broke it?" They ask, "How did our testing process miss this?" They seek the information required to fix the system.

This connects directly to the DORA research, which found that a generative culture predicts software delivery performance and organizational performance.[7]

In my experience, scaling organizations often aim for a generative culture but drift toward bureaucracy over time.

The Hero Trap vs. System Resilience

The long-tenured high performers I described earlier were not 'Heroes' in this toxic sense. They were force multipliers. They didn't just fix the legacy server; they taught the junior engineers how to fix it.

In the absence of a generative system, organizations rely on Heroes.

We all know the archetype. The "Brent" (from *The Phoenix Project*).[8] The senior engineer who is the only one who knows how to restart the legacy server. The manager who works every weekend to brute-force a release over the finish line.

For years, I made the mistake of rewarding this behavior. I praised the "extra mile." I gave awards for "saving the launch."

I was wrong.

A hero is a symptom of a system failure.

If success depends on one person rescuing the release, the process or architecture is already broken.

Relying on heroes creates a "Pathological" dependency. The hero becomes a bottleneck. They hoard knowledge (unintentionally or intentionally) because their value is tied to being the savior. And eventually, they burn out.

We saw a version of this on one of our value stream teams. We had a senior engineer with deep expertise who naturally became the go-to reviewer and design voice. He carried a disproportionate share of pull request reviews and was frequently the safety net when something felt risky or unclear.

From the outside, things looked efficient, but the model was more brittle than we realized. When he went on parental leave, throughput noticeably slowed, and our deployment cadence dipped.

It wasn't a crisis, but it revealed the real issue: too much system knowledge and decision-making had centralized in one person, and the rest of the team needed more shared ownership to keep momentum steady.

The needed shift was cultural: stop celebrating the saves and start valuing the boring releases.

The automated test suite that caught a bug before it reached production became worth celebrating because it meant no one had to be a hero on Friday night. The runbook that enabled a junior engineer to handle an incident without waking the tech lead became worth celebrating too. Praise shifted from individual rescue efforts to system resilience.

This transition is difficult. Heroes *like* being heroes. It feels good to be needed. Moving to a model of collective ownership requires strong leadership to reassure high performers that their value lies in mentoring and enabling, not just in coding and rescuing.

THE SCALING TRAP: EVOLUTION, NOT INSTALLATION

Changing culture, specifically moving from "Hero-based" to "System-based", is dangerous if you try to do it everywhere at once.

A common failure mode in transformation is the "Big Bang" rollout: announcing that on Monday, all 50 teams will switch to a new operating model. This is, in effect, a denial-of-service attack on your own organization.

To avoid this, I've found Jon Smart's mantra useful: Think Big, Start Small, Scale Out."[9]

- *Think Big:* Have a clear vision of the desired end state (e.g., "All teams operate as autonomous value streams with Generative culture").
- *Start Small:* Do not reorganize the division. Pick *one* pilot team. Give them the permissions, tools, and executive sponsorship to work differently. Treat this team not as a special project, but as a laboratory.
- *Scale Out:* Once that pilot team demonstrates better flow, higher morale, tangible outcomes, use their success as social proof. Invite a second team to adopt their patterns.

Transformation spreads through visible proof, not presentation decks.

Psychological Safety in the Real World

If you want to kill the hero culture and move toward Westrum's Generative model, the prerequisite is psychological safety.

Amy Edmondson defines it as "a belief that one will not be punished or humiliated for speaking up with ideas, questions, concerns, or mistakes."[10]

This is often misunderstood as "being nice" or avoiding conflict. In my experience, high-performing teams are rarely "polite" in the traditional sense.

They argue. They debate. They push back.

Psychological safety means the conflict is about the problem, not the person.

The difference becomes obvious when you compare two teams.

Team A (High Safety): This team operated at an elite level. They had an experienced agile delivery manager who acted as a coach. When the deadline shifted, they didn't panic. They got into a room (or a Zoom call) and adjusted the plan together.

The conversations often sounded like this:

- The engineer would warn, "If we rush this API, we introduce a security risk."

- The Product Manager would ask, "Okay, can we flag-gate it and release it internally first?"
- The QA lead would clarify, "I need two days for regression."

They debated trade-offs openly. No one was hiding the truth.

Team B (Low Safety): This team had a Product Manager who operated with a command-and-control style. Deadlines were treated as moral obligations.

- When QA raised a defect, it was often marked "deferred" to hit the date.
- When engineers raised technical debt concerns, they were told "we'll do it next sprint" (which never came).
- In meetings, the team was quiet. They nodded. They agreed.

Team B *looked* compliant. But their outcomes were poor. Defects leaked into production. Technical debt strangled their velocity.

The difference wasn't talent. It was safety. In Team B, speaking up was futile, so they stopped doing it.

DEVELOPER EXPERIENCE AS A BUSINESS DIFFERENTIATOR

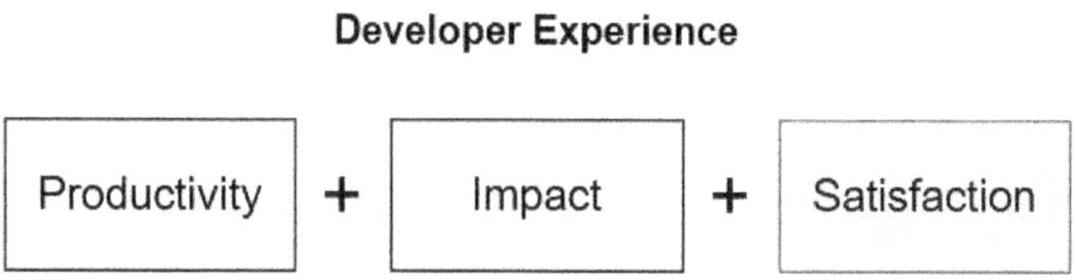

Figure 8.1: *The Developer Experience Equation.* DevEx is not merely about satisfaction; it is the sum of productivity, impact, and satisfaction.

Too many leaders view Developer Experience as a "perk," something akin to free snacks or ping pong tables.

In reality, DevEx can be a leading indicator of business performance.

McKinsey's Developer Velocity research reports that top-quartile software delivery organizations significantly outperform peers on business outcomes, including revenue growth.[11] GitHub's *Octoverse* report highlighted that teams with stronger automation and reduced friction were able to dedicate significantly more time to innovation.[12]

Recent research, including the *DX Core 4* framework, frames developer experience through feedback loops, cognitive load, and flow state.[13]

- *Feedback Loops:* How quickly can I tell whether my code worked?
- *Cognitive Load:* How much complexity and process overhead must I carry to do the work well?
- *Flow State:* How often can I make meaningful progress without interruption?

Sentiment tracking helped measure this. The questions were simple: 'Do you have the tools you need?' and 'Is the process helping or hurting?"

When we saw sentiment dip in a specific value stream, it was almost always a precursor to a dip in *Flow Velocity* and *Realization.*

Measuring developer experience is only half the work. The other half is communicating to developers what changed as a result of their feedback. When teams see that their input resulted in visible improvements, it reinforces the feedback loop and builds the trust that makes future surveys meaningful rather than performative.

Developer experience was treated as a core outcome of Platform as a Product. The Platform Team,[14] discussed in Chapter 5, was accountable for reducing friction for the stream-aligned teams. A slow build server was a defect in the platform product. A deployment process that depended on a manual approval ticket and added two days of delay was a blocker that needed to be removed.

Sustainability is the other side of the DevEx coin. Burnout is the enemy of Flow. Teams worked to build a sustainable pace, limiting Work In Progress (WIP) and treating incident reviews as learning opportunities rather than blame sessions.

Reality Check: Perfection was never the claim. Legacy pockets still existed where the "inner loop" from code to test was slower than it needed to be. But measuring DevEx alongside business metrics helped ensure the engine was not being burned out to win the race.

Navigating the Dip: The Engineering of Change

A "people-first" mindset means recognizing that people, not tools, are at the center of change. I often advocate the mantra "Think Big, Start Small, Scale Out," but we rarely discuss the friction required to get there.

Any new technology or process—whether it's a platform migration or the rollout of AI tools and agents—is disruptive. Even changes meant to improve DevEx and team efficiency will initially degrade it.

This is the J-curve of change: performance dips as teams unlearn old habits and build new ones. Modern organizations do not go through this once. They experience repeated waves of disruption and recovery, which over time look more like an S-curve.

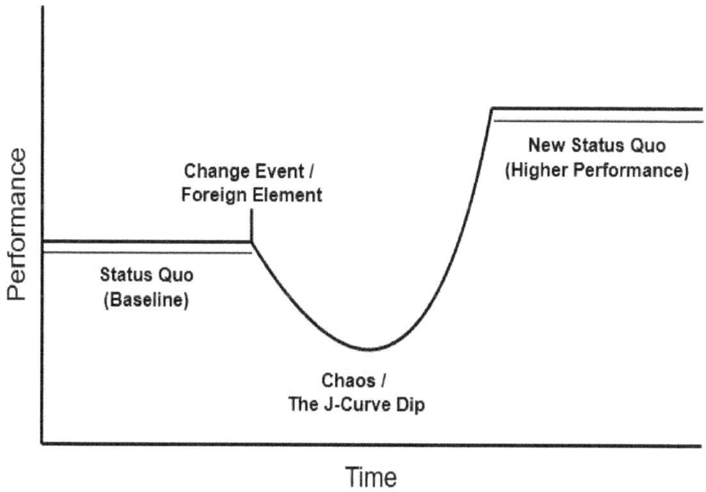

Figure 8.2: *The Change J-Curve.* Each change introduces an initial performance dip before a higher plateau emerges; over time, repeated J-Curves form an S-Curve pattern of stabilize, dip, stabilize higher.

The Stabilizing Force of Long-Lived Teams. This is why change has a higher success rate when anchored in long-lived, stable teams.

When a team has worked together for a long time, they possess a reservoir of trust and psychological safety. They can lean on established relationships while struggling with new practices or tools. If you try to force complex change on a newly formed or "fungible" team, the combined stress of "Storming" (from the Tuckman model) and learning a new workflow will break them.

To manage the J-Curve, we use a pilot strategy:

- *Start Small (The Pilot):* Begin with one or two high-trust, long-lived teams. Do not roll out to the whole org at once.

- *Concentrated support:* Give these pilot teams dedicated training, cohort collaboration channels, and direct access to experts.
- *Prove the Path:* Let these teams struggle, learn, and eventually climb out of the dip. Once they succeed, they become the social proof that convinces the rest of the organization.

Support is not optional. It is structural. As leaders, we cannot simply announce a change and expect flow. This aligns directly with the Discipline of "People Engagement" we defined in Chapter 6. We have to bridge the gap.

- *Acknowledge the Friction:* Communicate clearly that things will get bumpier before they smooth out. This honesty builds trust and prevents the "this isn't working" panic during the dip.
- *Surge Support:* During transition periods, we must offer dedicated office hours and documented migration paths. You cannot ask a team to maintain delivery velocity *and* learn a new paradigm without providing slack.

Peer-led Adoption (Scaling Trust)

You cannot scale change through executive mandates alone. Mandates drive compliance; they do not drive belief.

As John Kotter argued, sustainable transformation requires what he called a 'volunteer army': respected peers inside the teams who help others adopt change through credibility, coaching, and example.[15]

Trust travels faster horizontally, from engineer to engineer, than it does vertically from the C-suite. Change spreads through trusted peers inside the teams, not through title alone.

These pilot teams need trusted peers with deeper training who can help others translate new tools into real team workflows. That matters even more in AI adoption, where the shift is not just tooling but capability.

The Listening Loop

Finally, we must keep the feedback channels wide open. As teams implement change, new obstacles will emerge that our planning missed.

- *Local Ask Me Anything (AMAs) and Office Hours:* Large town halls often silence dissent. Local champions can hold smaller, "safe space" AMAs where developers feel comfortable sharing honest friction. Furthermore, teams that have successfully navigated the change can hold office hours to coach others.
- *Feedback to Action:* When a champion surfaces a blocker, leadership must act on it visibly. This closes the loop and proves that the "People First" value is real.

But listening requires patience. In the absence of trust or under extreme pressure, some leaders look for shortcuts. They stop asking the team how things are going and start asking the tools. They attempt to bypass human conversation entirely to find "objective" truth in the data.

When Metrics Become Surveillance

This brings us to the dark side of measurement.

The dark side of measurement appears when system metrics are used to judge individuals.

One request I've seen surface in multiple organizations is the "lines-of-code leaderboard." It sounds intuitive: more code should mean more productivity. But this metric destroys the behavior it aims to improve.

When metrics are used for surveillance, three things happen:

1. *Trust evaporates:* The team sees the tool as a weapon.
2. *Gaming begins:* Developers will write verbose code to pump up volume, or split one story into ten micro-tasks to increase "throughput."
3. *Outcomes suffer:* The focus shifts from "delivering value" to "feeding the metric."

Once a metric becomes a target under pressure, behavior bends toward gaming the number rather than improving the system.

Leadership Habit: We established a rule: metrics belong to the team.

- The *team* reviews their cycle time during the retrospective.

- The *team* decides if their WIP is too high.
- Management looks at the *aggregate* trends to ask: "How can I help remove the bottleneck?"

When I review value stream metrics, I am not looking for someone to blame. I am looking for the constraint in the system that needs to be removed. The moment a leader uses data to point a finger at a person, the culture begins to slide from Generative back to Pathological.

PAPER CULTURE VS. LIVED CULTURE

Paper Culture is what an organization says about itself.

Lived Culture is what shows up in decisions, tradeoffs, and behavior under pressure.

Every company has a "Paper Culture." These are the values written on the website, painted on the lobby wall, and printed in the employee handbook.

Integrity. Innovation. Excellence.

But *Lived Culture* is what happens when the server crashes at 4:00 PM on a Friday.

- Does the manager ask "How can I help?" or "Why isn't this fixed?"
- Does the team swarm to solve it, or do they look for someone to blame?
- Does the post-mortem focus on the root cause, or on finding a scapegoat?

I've seen this disconnect during acquisition integrations. On paper, an acquired company's values may resemble those of the acquirer. But integration pressure has a way of exposing the difference between what an organization says and what it does.

Under tight timelines and shifting priorities, collaboration became more cautious, fewer risks were raised early, and some decisions started funneling upward. This was not out of malice, but out of uncertainty and self-protection.

Rebuilding trust wasn't instantaneous. It required consistent leadership behavior over weeks, not slogans in a deck. The focus was on shielding teams from noise, making tradeoffs explicit, and turning slips into learning loops rather than blame cycles. When the deadline moved, the question wasn't "who failed?" It was "what did we learn, and what do we change?" Over time, people relaxed, candor returned, and the values stopped being "paper culture" and became the lived operating system again.

The Modern Agile Shield

I saw the inverse of this years later, during a later organizational transition.

Back in 2018, we had moved beyond generic corporate values and adopted the principles of Modern Agile as our cultural operating system. We anchored our work in four active behaviors: Make People Awesome, Make Safety a Prerequisite, Experiment & Learn Rapidly, and Deliver Value Continuously.[16]

These weren't posters on a wall; they became the way we worked day to day.

During organizational transitions, there's always pressure to standardize culture: adopt the new language, align the rituals, and let distinct operating principles fade in the name of "alignment."

The division took a different path, aligning with the broader mission while holding fast to Modern Agile as its local dialect.

As new controls and management changes introduced pressure, the leadership task was to absorb it without sacrificing continuity, ownership, and trust. Because those principles were lived daily, they became a shield. a shared vocabulary to defend high-performance practices without sounding obstructionist.

Over time, it became possible to work effectively inside a larger system while protecting the operating model that made the teams successful in the first place.

Reflection: A colleague once told me, "You don't see a football team's culture when they're winning 40–0. You see it when they're losing in the fourth quarter. Do they turn on each other? Or do they execute the next play?"

That's the difference: paper culture is what you say. Lived culture is what you do when it costs you something.

PUTTING IT INTO PRACTICE

Investing in culture is not about buying more ping pong tables. It is about rigorous engineering of the social system.

1. Diagnose Your Westrum State

- Look at your last major incident. Was information sought (Generative) or hidden (Pathological)?
- Look at your cross-team dependencies. Are they conversations (Generative) or tickets thrown over a wall (Bureaucratic)?

2. Kill the Hero

- Identify the "Brents" in your organization.
- Task them with documenting and automating their knowledge.
- Make their primary goal *enabling others*, not *doing the work*.

3. Measure Sentiment as a Leading Indicator

- Don't wait for exit interviews.
- Use brief, frequent surveys (e.g., "I feel I can deploy software safely").
- If sentiment drops, treat it like a production outage. Investigate and fix.

4. Shield the Team

- Your job as a leader is to be the heat shield. Absorb the chaos.
- Pass down *context*, not *stress*.

KEY TAKEAWAYS

- *Culture is a constraint.* A technology organization cannot scale on fear, bureaucracy, or heroics; speed depends on trust, candor, and shared ownership.
- *Psychological safety precedes performance.* When people fear speaking up, risks stay hidden, metrics get gamed, and learning slows.
- *Heroics are not resilience.* Strong systems spread knowledge, reduce single points of failure, and reward enablement over rescue.
- *Developer experience is not a perk.* It can be an early indicator of delivery health, and stronger feedback loops, lower cognitive load, and less friction support better outcomes.
- *The leadership task is practical:* protect one value stream, make tradeoffs explicit, reduce decision latency, and treat AI as capability-building rather than tool deployment.

PART 2 RECAP: THE MACHINERY OF TRANSFORMATION

Part 2 moved from mindset to mechanics. The shift was from ideas to infrastructure: how leaders build systems that make modern practices durable under pressure.

What Changed in the Model

1. *Structure (Ch. 5):* The model moved from projects to products to value streams, protecting flow rather than fragmenting it.
2. *Funding + Governance (Ch. 5):* Funding shifted from temporary scope to sustained product investment through stable teams, replacing annual resets and milestone gates with ongoing steering and outcome-based review.
3. *Future-proofing (Ch. 5):* AI was treated as capital and capability-building, not as a typing upgrade, so it strengthens healthy systems instead of accelerating dysfunction.
4. *Behavior (Ch. 6):* Leadership reflexes gave way to discipline, explicit priorities, decision rights, and protection of the system under pressure.

5. *Navigation (Ch. 7):* Strategy theater was replaced by alignment around shared objectives and unified signals, connecting work to business intent rather than local priorities.
6. *Culture (Ch. 8):* Trust, psychological safety, and DevEx became foundational because the human system determines whether the operating model is real or merely declared.

Durability shows up when teams stay stable longer, decision latency drops, tradeoffs become visible, and metrics are used to learn rather than to police.

The Pivot Ahead

Structure, leadership, alignment, and culture create the foundation for transformation. But the next executive question is not cultural; it is economic. How do these choices show up in outcomes, risk reduction, and return on investment? Part 3 turns from operating confidence to business proof, tracing how engineering investment becomes measurable value.

PART 3: THE VALUE STREAM EXECUTIVE

Linking Technology to Business Outcomes

9

The Business-Technology Connection

"Productivity is meaningless unless you know what your goal is." — Eliyahu M. Goldratt, The Goal[1]

Technology teams often know they are making the business better long before the business can describe it. The software ships, the customer experience improves, risk is reduced, and new opportunities open up, yet technology is still too often viewed through the narrow lens of spend. When leaders cannot clearly see how investment translates into results, engineering is managed like a cost center rather than a strategic partner.

Culture gives people a clear sense of purpose and priorities. Connection gives them proof that their work matters. But here is what neither can overcome on its own: invisibility.

In Part 2, I explored how trust and leadership create the conditions for alignment around shared goals and outcomes. But alignment alone is not enough if the work itself remains hidden. Productivity becomes easy to misread when the goal is vague, the path is unclear, and the results are hard to trace.

This chapter focuses on making that connection visible. It examines four elements that help close the gap between effort and impact: value, flow, realization, and the financial view of the value stream that connects investment to outcomes. Together, they give leaders a practical way to make invisible work visible and connect technology to business results.

THE INVISIBLE FACTORY FLOOR

Software engineering has always been expensive. Production support, infrastructure, database engineering, and the scaffolding that keeps systems running in the background all come with a high price tag.

Executives have never missed the cost. What they missed was the line between value and cost.

Unlike a factory floor, there was nothing visible to watch. Leaders could not see raw materials moving into finished goods. They only saw a large headcount line in the budget and an endless backlog of requests. The work was invisible. The investments were abstract.

In that environment, the business had no choice but to "trust" technology leadership. When systems remained stable, that trust was maintained. When incidents flared, skepticism returned, and technology was managed as overhead.

That invisibility has always been our curse.

This is *Funding Reset Gravity* in action, one of the Five Transformation Friction Triggers, where economic pressure quietly pulls organizations back into project-era behavior.

I know this firsthand. Too many times in my career, I have sat on the leadership side of the table fighting to defend roles and names, while also knowing that executives above me would expect specific numbers from us. Engineering salaries were among the first targets when financial pressure mounted. Reductions in force were the easiest lever to pull.

But those headcount equations rarely accounted for the work that would never be delivered or the strategic bets that would quietly stall.

I have acted on those decisions, and I have seen the aftermath: projects delayed, morale damaged, trust eroded. The reflex is always short-term. Cut expenses today, and only later come to grips with the deeper cost of what was lost.

When external partners are part of the delivery system, the business–technology connection depends on governing Flow and Realization end-to-end, a challenge explored further in Appendix B: *Profitable Engineering in a Vendor-Heavy World.* When leaders can't see value, they manage cost, and cost management always finds engineering first.

Cost discipline matters. The risk begins when cost-cutting becomes detached from value, capacity, and long-term consequences. But when the work remains invisible, leaders fall back on the categories and controls they inherited. That is where cost-center thinking returns.

Why Cost-Center Thinking Still Reappears

Technology is often treated as overhead because modern organizations inherited financial models built for a different era.

Modern finance emerged in an industrial economy that prioritized cost accounting above all else. Work was divided into cost centers. Expenses were allocated with precision. Efficiency became the dominant concern. Value was assumed to come from markets or operations, not from the functions being tracked.

When IT entered the picture, it fell neatly into that framework. The residue remains stubborn.

Budgets are still controlled by finance. Even when investments are made by product, they are usually tied to annual funding cycles. That cadence forces leaders to frame investments in terms of what will be delivered over the next 12 months, rather than continuous returns over the product's life. Governance checks deliverables more often than they check whether business outcomes actually change.

Financial treatment can add another layer of distortion. In many organizations, the way technology spending is categorized and reviewed still favors short-term cost visibility over long-term capability. Work that improves the product, strengthens the platform, or expands future options can still show up primarily as an expense. The result is predictable: technology gets discussed as overhead even when it is building capability that the business will rely on later. Without a new way to measure and fund, the business–technology connection remains obscured.

A LANGUAGE EXECUTIVES CAN ACTUALLY HEAR

I learned early on that dashboards are only helpful if they change decisions. The breakthrough came when I stopped trying to communicate a new vocabulary to executives and started framing the conversation in language they already used.

The first part of that language was value: what changed in growth or efficiency, expressed through a small set of product-specific signals. The second part was flow: how reliably the system turned investment into value.[3] The third part was realization: making outcomes visible in business terms.

Flow shows how efficiently work moves through the system, while realization captures how effectively those investments translate into tangible business outcomes.

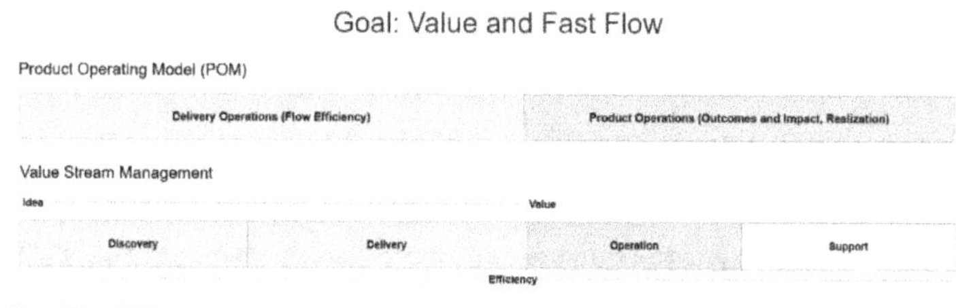

Figure 9.1: *From Flow to Realization.* To speak the language of business, we must connect operational efficiency (Flow) directly to market outcomes (Realization).

THE PARTNERSHIP CONTRACT

A strategic partnership between Product and Engineering is not a recurring meeting. It is a shared operating agreement.

Product and Engineering co-own outcomes. They also co-own constraints: reliability, compliance, architecture health, and cost-to-serve.

Finance co-owns the scoreboard: what can be capitalized, what must be expensed, and how investments are reflected on the balance sheet.

The partnership becomes real when "done" means realized, not merely shipped.

In practical terms, the partnership must produce four shared artifacts. This partnership is where EBITDA (Earnings Before Interest, Taxes, Depreciation, and Amortization) is protected or diluted because it determines what gets funded, what gets fixed, and what becomes invisible tax.

- **Outcome Bets:** A small set of measurable bets tied to customer behavior, revenue expansion, efficiency gains, or risk reduction.
- **Roadmap sequencing with explicit tradeoffs:** Not just what you'll do—what you will *stop*, and why.

- **Shared risk register:** Architecture fragility, compliance exposure, data quality, AI governance, and operational bottlenecks made visible and reviewed with cadence.
- **Decision log:** The tradeoffs, assumptions, and economic implications written in plain language so leadership builds institutional memory.

When these artifacts exist and are reviewed with cadence, the business–technology connection moves from philosophical alignment to operational discipline.

Without them, the partnership stays aspirational. With them, it becomes an engine of realization that turns decisions into visible motion instead of leaving them stranded in documents or intentions. When strategy is managed as a living dashboard rather than a static slide deck, momentum compounds, trust rises, and outcomes no longer depend on heroics.

When we told the story this way, through value, flow, and realization, the conversation shifted. Leaders no longer had to triangulate from cost and output to guess at impact. They could see enough to ask the right next question. Value, flow, and realization are not just engineering signals. They are key indicators of overall system performance. They are the bridge between technology and the boardroom.

Funding Durable Teams and Governing Outcomes

If you want accountability for outcomes, you must align funding accordingly.

Project Funding (Short-term scope)
↓
Annual Product Funding (Annual deliverables)
↓
Continuous Product Investment (Ongoing ROI across product life)

It is difficult to ask a team to own an outcome while funding it as if it were a temporary project.

That model made sense in an era when most organizations were structured around functional teams, time-bound initiatives, and annual delivery commitments. But I have seen, including in my own environment, how quickly it breaks down when leaders expect durable ownership, continuous learning, and measurable outcomes. It looks coherent on a slide and falls apart in practice.

"Outcomes over outputs" is easy to say and much harder to fund. In many organizations, annual budgets, milestone governance, and planning cycles still reward scope and delivery dates more than sustained business returns. Even where Agile and DevOps practices have taken hold, the funding model often pulls teams back toward one-year commitments, fixed dates, and recurring resets.

I saw that in my own environment. Annual product budgets were a meaningful step forward from traditional project funding, but they still stopped short of true long-term product investment.

The times we came closest to something better were the moments when teams had stable budgets, leadership support, and enough continuity to stay focused.

When funding and accountability were stable, teams began to say "not now" to distractions. Leaders began to expect a steady narrative of progress instead of a sequence of annual resets.

The pivot requires four moves:

- Make value flow visible.
- Build durable teams rather than temporary projects.
- Define outcomes at the product level and tie them to company goals.
- Instrument the work so progress can be reviewed regularly and adjusted.

None of these is theoretical. They are the mechanisms by which organizations shift from reporting activity to demonstrating impact. The goal is not simply to fund work differently. It is to create a system in which technology can be governed as an ongoing investment with visible returns rather than a recurring budget request.

Annual product budgets are a meaningful step forward compared with project-based funding, but they still limit flexibility. Treating products as ongoing investments, sustained through stable teams and reviewed against outcomes rather than reset each year, is what ultimately unlocks accountability. Over time, that is how returns compound across the life of a product.

Why it matters: This is the bridge between the engineering discipline and the CFO's worldview. Profitability by stream turns "trust us" into "here's the return."

THE VALUE STREAM P&L

A product operating model needs a financial spine. Otherwise, we rename projects while keeping the same incentives.

That is why I recommend operating with a Value Stream P&L (Profit & Loss): a simple, repeatable way to connect stream-level investment to measurable outcomes. It does not need to be perfect to be useful, but it must be consistent enough to guide decisions.

This works when the organization funds products and value streams rather than temporary projects. A product operating model treats each value stream as a long-lived investment, not a one-time initiative. Teams become durable units aligned to a product or domain, and leadership evaluates that investment over time in terms of return, not just delivery. The Value Stream P&L provides the financial lens for that conversation, making it possible to review investment, flow, and outcomes together.

One of the most important advances in modern software delivery has been the ability to make flow visible. With DevOps Research and Assessment (DORA) metrics, Software Engineering Intelligence platforms, and Value Stream Management tools, organizations can now understand how work moves from idea to production, where it slows, how much is in progress, and how efficiently effort becomes deployed software.

That shift has been transformative. For most of software delivery's history, leaders could see the budget, but not the path the work took through the system. Now they can answer questions that once felt impossible: how long delivery takes, where bottlenecks form, and how capacity is really being used.

But even as visibility into flow has improved dramatically, one critical question remains unanswered in many organizations.

Did the work actually produce value?

Measuring how efficiently software moves through the system is important. But efficiency alone does not guarantee impact. A team can deliver features quickly that no one uses. A platform group can ship infrastructure improvements that customers never notice. A product team can deliver a roadmap on schedule without meaningfully improving the customer's experience.

This is the gap between flow and realization.

Flow tells us how efficiently work moves through the system. Realization tells us whether the work produced the outcomes we intended.

Closing that loop is essential if technology organizations want to operate as true business partners rather than cost centers.

This is where the Value Stream P&L becomes useful. In most organizations, the cost of software production is not a mystery. Salaries, tools, cloud spend, and vendor costs can usually be estimated with reasonable confidence.

A Value Stream P&L also requires technology leaders to understand both the cost structure and the product or value stream's margin profile. In a product operating model, they should know what it costs to build and run the product, what margin it generates today, and what margin it is expected to generate over time.

That includes the ongoing cost of development, production operations, and support. The goal is not perfect precision. It is enough economic clarity to judge whether the stream can sustainably fund the team and platform that support it. If the cost to build and run the product consistently outweighs the value it returns, leaders need to know why, for how long, and what must change to justify the investment.

The harder question is whether that investment produced value. Flow shows how work moved; realization shows whether it changed something that mattered.

The figure below captures that loop. When leaders review cost, flow, and realization together, they can see not just what software costs to build, but what the investment actually returns. That is the essence of the Value Stream P&L.

That connection does not require perfect financial attribution. The link between engineering work and revenue is often complex, but organizations can still build meaningful signals that connect delivery to outcomes.

Those signals may appear in customer behavior, product usage, or business performance: adoption rises, friction falls, retention improves, or support costs decline. Reviewed together, flow and realization shift the conversation from activity to impact by showing both how work moved and whether it produced meaningful results.

Some larger organizations support this discipline through a Product Operations (Product Ops) function. Product Ops does not replace Product or Engineering ownership. Still, it can help maintain the data, process consistency, and cross-functional coordination required to define anticipated outcomes, measure actual outcomes after release, and keep follow-through from fragmenting as the organization scales. Atlassian and other product-management sources describe Product Ops as the function that helps product organizations scale by managing data, workflows, and operational consistency across teams.

That shift changes the nature of software delivery. Traditional project thinking treats delivery as a linear process: idea, build, ship, done. Modern product organizations operate differently. Delivery becomes part of a continuous learning system.

After the software is released, the most important phase begins: observing what actually happens. Do customers use the feature? Does behavior change? Did the improvement solve the problem it was intended to solve?

Those observations feed the next decision.

The system becomes a loop rather than a line.

Idea → Design → Build → Deploy → Operate → Outcome → Learn → Improve

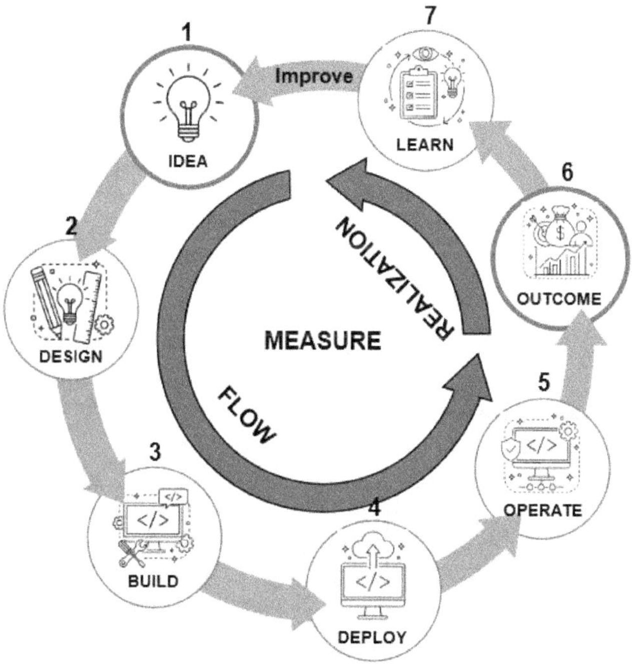

Figure 9.2: Closing the Loop from Flow to Realization

Delivery is not the end of the value stream. Realization and learning complete the loop.

This is where value stream management, product thinking, and continuous improvement converge. The value stream is not just a pipeline for delivering work. It is a learning engine for producing better outcomes over time.

For engineering leaders, this shift has profound implications. When organizations measure only delivery efficiency, engineering becomes a function focused on throughput. When organizations measure both flow and realization, engineering becomes a strategic contributor to business performance.

Teams begin to understand not only what they are building, but why it matters. They can see the relationship between their work and the value it creates for customers and the business.

Over time, that visibility transforms the conversation between product, engineering, and executive leadership. The question is no longer *How much software did we deliver?*

Instead, the conversation becomes:

What outcomes did that work produce, and how can we improve the system to produce even better results?

That is the essence of the Value Stream P&L.

REAL-WORLD APPLICATION:

The Due Diligence Test

I've been on the technical side of multiple acquisitions and major investment rounds. In those moments, investors and acquirers are not asking about Jira velocity or sprint burndowns.

They ask about risk and scalability:

- "Can this platform support 10x growth?"
- "How much technical debt are we buying?"
- "How quickly can we integrate this into our portfolio?"
- "Is the culture a flight risk?"
- "How quickly can we deliver value?"

Serving alongside colleagues in technical diligence during those funding rounds taught me that silence or vague answers can devalue a deal. You need data. When we answered with Flow efficiency, quality metrics, and evidence of a scalable, autonomous team structure, we weren't just selling software. We were showing the operational evidence behind the valuation.

A Short Story About Tradeoffs

Picture a quarterly planning session.

Product brings forward a feature expected to improve conversion in a key workflow. The revenue case is attractive, Sales sees upside, and the CEO wants momentum.

Engineering raises a different signal. The workflow runs on aging architecture. Incident frequency is rising, lead time has lengthened over the past two quarters, and defect escape is up. The system is still working, but the margin for error is shrinking.

Finance adds another constraint. Refactoring will land in operating expense. EBITDA pressure this quarter is real.

Three paths emerge:

1. Ship the feature immediately and defer architectural work.
2. Refactor first and delay the revenue opportunity.
3. Sequence both intentionally. Limit scope, address the highest-risk constraint, and instrument the release for measurable realization.

In a traditional cost-center environment, option one usually wins. Short-term optics dominate. Velocity appears intact—until rework, incident load, and support costs rise.

In a mature partnership, the conversation shifts.

Engineering presents flow signals: lead time trends, WIP aging, and defect ratios. Product clarifies the expected behavior change and defines measurable success criteria. Finance reviews the Value Stream P&L and sees that incident cost and support load have already increased cost-to-serve.

The question becomes different.

Not: "Can we afford to refactor?"

But: "What is the economic cost of not stabilizing the system before increasing load?"

The team chooses the third path.

They narrow the feature scope. They address the architectural constraint most responsible for defect escape. They define success criteria before writing code. They instrument the release.

Three months later, the feature delivers a measurable improvement in conversion rates. Incident load stabilizes. Lead time contracts. Support volume declines.

Refactoring did not slow velocity. It preserved it.

That meeting is the business–technology connection in action.

VALUE STREAM MAPPING AND ROADMAPS THAT EARN TRUST

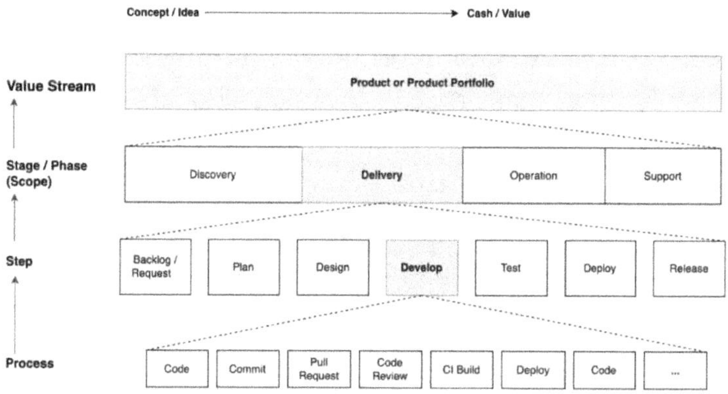

Figure 9.3: *The Hierarchy of Flow.* Mapping must occur at multiple levels of granularity, connecting the high-level Value Stream down to the specific Process step.

During that period at Parchment, I participated in outcome and current state mapping sessions with one of the more mature Agile teams.

The work made a recurring source of friction visible: shifting requests and midstream priority changes were slowing delivery and draining focus in ways that conversation alone had not fully exposed.

Leaders could finally see the cost of constant reprioritization and begin discussing where intervention was actually needed.

But the harder problem was never producing the map once. It was sustaining the practice. Without cadence, alignment, and executive sponsorship, mapping never became muscle memory.

That is a common pattern in Value Stream Management efforts: the workshop happens, the insight is real, and then the rhythm disappears. The lesson is that mapping must be treated as a living practice, not a one-off event.

The same was true for Flow Roadmaps. The next intended step was to turn those insights into a visible portfolio of system improvements tied to outcomes, ownership, and follow-through. But without consistent buy-in across product and leadership, the practice remained aspirational.

That experience reinforced a larger truth: these tools do not survive inside engineering alone. They become powerful only when they are shared across roles and reviewed with cadence.

The artifact only becomes valuable when it is supported by follow-through, shared accountability, and a recurring cadence of review.

The difference between theater and transformation is leadership buy-in, trust, and that sustained rhythm of use. Without those conditions, maps and roadmaps remain static documents. With them, they become the negotiation surface for investment, tradeoffs, and coordinated action.

Why it matters: mapping turns friction into evidence, and evidence turns "do more with less" into "fix this constraint first."

FROM FEATURE FACTORY TO OUTCOME ENGINE

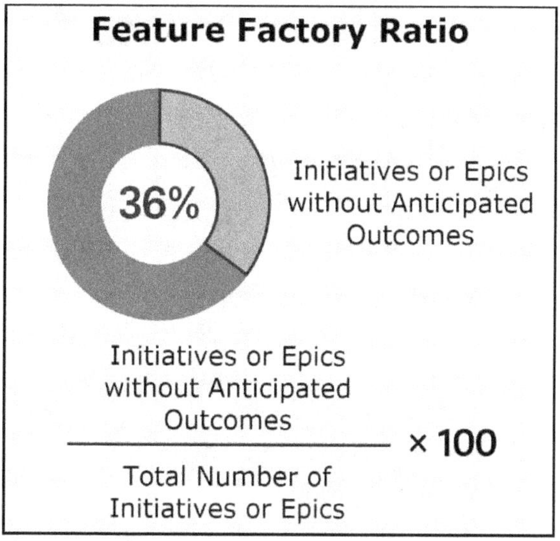

Figure 9.4: *The Feature Factory Ratio.* The share of initiatives with clearly defined anticipated outcomes versus work delivered without explicit outcome criteria.

The phrase "Feature Factory" entered our vocabulary when we realized how easy it is to ship constantly and learn almost nothing. The deeper issue was not speed itself, but the lack of a feedback loop. In the strongest moments, we defined success before writing code, instrumented the release, and reviewed what actually changed afterward.

Even where that discipline was uneven, the result was visible: less wasted effort and more trust from leadership.

One way I tracked that discipline was through a simple ratio I created, the Feature Factory Ratio. In our Jira-based delivery practice, every initiative and epic was expected to include an anticipated outcome before work began.

That made it possible to look back and see how much work entered the system with a defined outcome versus how much did not. In software delivery, the Definition of Done is the shared understanding of what must be true before work is considered complete. It means more than writing code. The work has been tested, validated, and documented, and meets the standards required to release it with confidence.

I followed and promoted that discipline throughout the later part of my career because it created clarity, improved quality, and helped teams avoid calling something finished before it was truly ready.

The conversation needs to mature further.

It may help to separate the Definition of Delivered from the fuller Definition of Done.

Delivered means the work shipped. It moved through the system and reached production.

Done, in a broader and more meaningful sense, may come later, after the organization has learned from the result. Did the work produce the impact that was anticipated? Did it improve something that mattered for the customer, the business, or the system?

If that loop never closes, the work may have been delivered, but calling it done may be too generous. Release is an important milestone. Realization is what tells us whether the effort actually mattered.

The ratio did not prove whether those anticipated outcomes were ultimately realized. It showed whether the organization was building the habit of defining intended value before implementation started.

Closing the loop is not just good delivery practice; it is essential. It is what proves to the business that technology investment is producing measurable results, not just outputs. And once that loop becomes visible, another truth comes into focus: operating models and incentives determine whether teams optimize for outcomes or simply chase dates.

THE OPERATING MODEL MOVES THAT UNLOCK OUTCOMES

The operating model does more than organize work. It shapes behavior.

Early in the transformation, deadlines still shaped the system's behavior. Teams were optimized to hit dates, even when quality weakened, and the costs appeared later in rework, defects, and support burden. That is why the shifts below begin with funding, governance, and measurement.

The goal is not to reduce cost in isolation, but to build a system that improves speed, resilience, and outcomes together.

Table 9.1: Traditional IT Management vs. Value Stream Thinking

Traditional IT Management	Value Stream Thinking
Focus on cost and activity	Focus on value and outcomes
Project funding with milestones	Product funding with steady outcomes
Success measured by scope and dates	Success measured by changes in growth, efficiency, and risk

Siloed roles and handoffs	Cross-functional teams aligned to a value stream
Governance by status reports	Governance by decisions and constraint removal

And the most revealing shift is what gets measured—activity versus impact:

Table 9.2: Traditional IT Metrics vs. Outcome-Driven Metrics

Traditional IT Metrics	Outcome-Driven Metrics
Number of deployments	Measured change in retention, expansion, or conversion
Feature completion rates	Reduction in cost to serve on a target workflow
Sprint velocity	Competitive position or customer satisfaction tied to a change
Cost reduction totals	Value delivered to users and the economics attached to it

When Product and Technology operate as a single system, technology stops being a cost center and becomes a multiplier on revenue, margin, and valuation. A high-performing partnership changes unit economics:

- Cost-to-serve drops as toil is removed through automation, fewer incidents, faster recovery, less manual QA, and fewer escalations.

- Revenue becomes more durable as reliability, quality, and trust improve retention and expansion.
- Growth accelerates as cycle time compresses from insight to shipped outcome, and experimentation becomes cheaper.
- The risk discount shrinks as security, compliance, and operational fragility are addressed early rather than paid for later.

The inverse is also true. If the operating model produces hidden rework, fragile architecture, and reactive incident load, every $1 "saved" in capability becomes $10 in loss through churn risk, slowed roadmaps, support burden, and missed market windows. EBITDA doesn't forgive invisible tax.

PURPOSE AS A PERFORMANCE ADVANTAGE

When engineers can see how their work moves a customer metric, reduces risk, or strengthens the company's position, their work stops feeling like ticket completion and starts feeling like contribution.

Visibility does more than improve economics. It restores purpose. That sense of purpose strengthens retention, preserves context, and reinforces the conditions for better flow, stronger realization, and greater trust. In that environment, culture is no longer a soft topic. It is a structural advantage. The engineering organization becomes not only a delivery engine, but a multiplier of business capability and, over time, valuation.

Conclusion: Leadership Makes the Connection

I have worked in organizations that treated technology strictly as a cost center, and I have seen glimpses of what it looks like when technology is treated as the engine of strategy.

We are not fully there, but those glimpses along with the lessons from peers who have made it further are proof of what is possible when leaders choose to measure outcomes, fund for stability, and govern with flow in mind. When value, flow, realization, and financial visibility come together, technology investment becomes understandable in the same language the business already uses.

Good software makes the complex appear simple. Customers experience that simplicity only when leaders do the hard work of making the system around the code legible.

That is the job of leadership.

The business-technology connection becomes real only when it is practiced as a discipline visible in meetings, trade-offs, and evidence-based decisions. It is the point at which technology stops looking like overhead and starts showing up as a true business partner.

Key Takeaways

- *Speak the language of business:* credibility grows when technology is discussed in terms of value, flow, and realized outcomes rather than activity alone.
- *Visibility must change decisions:* maps, metrics, and review mechanisms matter only when they expose friction, sharpen tradeoffs, and create shared accountability.
- *Funding changes behavior:* durable teams and outcome-based governance outperform temporary project structures; continuous product investment is what compounds returns over the life of a product.
- *Incentives reveal priorities:* date-driven incentives buy compliance and hidden costs, while outcome-based incentives reinforce system health, learning, and long-term value.
- The business-technology connection must be made visible: when leaders can see how work, money, and outcomes connect, technology is far more likely to be treated as a strategic asset rather than overhead.

A dashboard that makes sense changes the tone of every conversation and creates a new obligation. If we can demonstrate how work becomes value, we must connect those signals to what boards and CFOs evaluate daily: revenue growth, margin expansion, cost-to-serve, and risk reduction. That is where software strategy and business strategy either move in lockstep—or not at all. Chapter 10 shows how to make that connection operational.

10
The Future of Engineering Leadership

"A leader is best when people barely know that he exists, … They will all say, 'We did this ourselves.'"[1]

As of late 2025, even the most seasoned technologists hesitated to predict the next few years. Public forecasts were increasingly bold, and some prominent leaders argued that AI's long-term impact was still being underestimated.[2]

Whether or not those timelines prove true, one certainty remains: the role of engineering leadership will evolve faster than any of us can script, demanding adaptability, ethics, and vision in equal measure.

I have watched entire teams redefine their value overnight—some thrive in the shift, others freeze in it. The difference rarely comes down to skill; it comes down to how leaders respond when certainty disappears.

WHAT CHANGES, WHAT DOESN'T

Amid all the change, it is worth asking: what changes and what doesn't?

There is a moment in every transformation when the dashboard goes quiet. No urgent pings. No red lights. No fire drills. Work is flowing, and yet leadership remains essential. Flow is not a natural state; it is a designed one, sustained by systems that evolve, leaders who clarify outcomes, and measurement that tells the truth.

Over the past few years, I have learned that the future of engineering leadership has less to do with directing tasks and more to do with designing environments.

The organizations that thrive will be those where leaders do not just supervise motion; they create the conditions for trust, safety, and velocity without shortcuts.

By late 2025, in the division I led, the shift toward AI-assisted work was still in its early stages. Team members had access to approved AI tools, but adoption depth varied. I did not see that as a failure. It was the natural phase of learning.

My job is not to sell people on the idea that AI will change everything. It is to help them see how it changes their work today, how we measure that impact, and how to prevent shortcuts from eroding the quality we depend on. That is where leadership earns its relevance.

FROM MANAGERS TO SYSTEM DESIGNERS

Old leadership models centralized decision-making at the top and measured progress through visible metrics, such as how many tasks were checked off and how many tickets were closed.

Modern leadership changes the emphasis. It is about creating systems where work can flow with less friction, learning can compound, and outcomes remain visible.

Leaders who thrive in this environment learn to read the system and change it. They focus less on counting activity and more on removing friction, building alignment, and making it safe to tell the truth about what is working.

In this new era, a leader's value is shaped less by how tightly work is controlled and more by how well the environment is designed. A traditional manager asks, "When will this be done?" A stronger leadership question is, "Where is the friction, and what investment would remove it?"

That shift changes how leaders think about budget, autonomy, and flow. The organization stops looking like a hierarchy of tasks to supervise and starts looking more like a system of value creation that needs clarity, trust, and fewer constraints.

When leaders miss this, AI becomes a gimmick instead of an amplifier, platforms become cost centers instead of accelerators, and metrics are gamed instead of trusted.

As defined earlier, every value stream is anchored to a long-lived product that owns outcomes end to end. That alignment gives leaders a more consistent way to reason about flow, funding, and customer impact. Strategy is no longer only what gets said in the boardroom. It is also expressed in how work is shaped, funded, and prioritized across the backlog.

Five Shifts You Cannot Ignore

Why it matters: These shifts aren't trend-watching. They're competitive economics. Leaders who adapt will compress time-to-value and reduce risk; leaders who don't will pay for the same work twice through rework, churn, and stalled execution.

Shift 1 – AI-Augmented Work Becomes the Default Across the Whole Stream

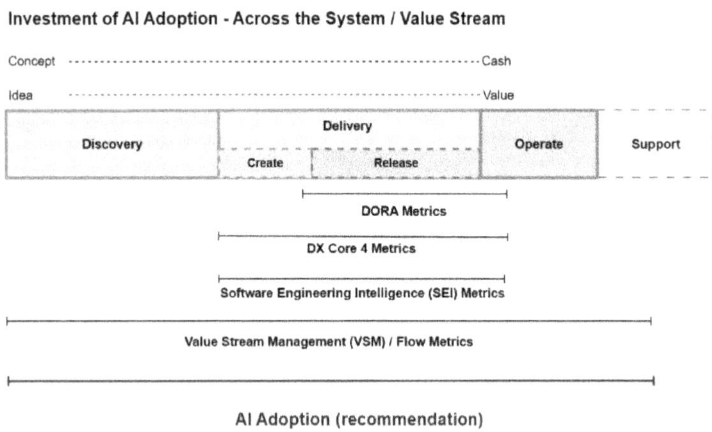

Figure 10.1: The AI-Augmented Value Stream. AI investment must expand beyond just Create (coding) to support the full stream (Ideate → Create → Release → Operate) measured through a hierarchy of metrics.

AI is often framed as a development accelerator such as cases with code generation, review, refactoring, and test scaffolding. That is useful, but if we only optimize that slice, we accelerate the smallest fraction of the system delay.

Most of the friction in software delivery lives far earlier and later in the flow.

Analysis across industries consistently shows that nearly half of the total delay occurs in *Ideate* and another large share in *Release*. At the same time, *Create*—the coding phase—is typically the smallest portion.[3]

The right move is to embed AI across the entire flow.

Ideate → *Create* → *Release* → *Operate*, and measure impact at the system level, not just at the keyboard.

What this looks like in practice:

- *Ideate:* Use AI to synthesize research, draft business cases, model risks, and surface queues or wait states before they cascade.
- *Create:* Apply AI to coding, reviews, and test generation, but tie results to outcomes and stability, not vanity speed-ups.
- *Release:* Automate compliance checks, dependency analysis, integration alignment, and deployment choreography to reduce handoffs.
- *Operate:* Detect incident patterns, suggest diagnostics, and automate low-risk remediation, then feed learnings back into product strategy.

The goal is to compress the idea-to-value cycle, not merely accelerate delivery. When AI adoption works, flow time shrinks, efficiency rises, waste drops, and stability holds or improves.

Why it matters: If AI only speeds up coding, you get local efficiency and global disappointment. The business wins only when AI reduces end-to-end decision latency and improves outcomes without increasing operational risk.

During that period, developers had access to approved AI tools, including *GitHub Copilot* and *Claude Code*. Some designers and UX practitioners were also piloting AI-assisted workflows.

The release process was already heavily automated across testing, validation, and deployment, with AI experimentation beginning to emerge there as well.

I did not measure success by license usage, prompt counts, or keystrokes. I anchored it in flow efficiency, defect escape, time to restore, and DevEx. Where AI improved those signals, the practice was expanded. Where it did not, the approach was adjusted.

Shift 2 – Flow and Outcomes Trump Output

For more than a decade, research has consistently shown that organizations that deliver quickly and reliably are far more likely to achieve their business goals.[4] The lesson is not that velocity wins; it is that system health wins.

Leaders who only measure output—tickets closed, points burned—push teams toward motion instead of meaning.

In practice, this means moving debates out of sprint ceremonies and into the design of the value stream. Where does work wait? Where does quality erode? Where are handoffs still occurring, and how can we eliminate them? When teams see the system rather than just their sprint, they stop arguing over points and start fixing constraints.

During that period, flow metrics were the backbone of my leadership. Teams measured flow time, efficiency, and load alongside outcomes such as adoption, retention, margin, and incident impact. DevEx surveys captured the human side. Velocity remained a planning signal, not a measure of success.

The Team's Dashboard: Flashlight, Not Hammer

There is a danger in discussing metrics at this level: it is easy to view them solely as a "report card" for executives. If teams believe flow metrics are being used to judge their personal performance, they will optimize for the metric, not the outcome. They will close tickets prematurely to improve "Flow Time" or avoid logging bugs to protect "Defect Density." The moment metrics become targets, they cease to be measures.

To prevent this, leadership must clarify the distinction between *reporting* and *feedback*.

For the team, metrics should function like a debugger for their process. They are the feedback loop that illuminates where the system is fighting them.

- *High Flow Load?* This is not a failure; it is a signal to stop starting new work and swarm to finish existing tickets.
- *Low Flow Efficiency?* This is not a sign of laziness; it is a signal that the team is blocked by external dependencies or waiting on slow reviews.
- *High Change Failure Rate?* This is a signal that the testing safety net has holes that need filling before we push for more speed.

When metrics are treated as feedback, they empower the team to run experiments. A team might say, "Our wait time on code reviews is killing our flow efficiency. Let's try a 'pairing hour' every morning for one sprint and see if the needle moves."

In this model, the team owns the dials. They use the data to construct the argument for what they need whether that is less WIP, better tools, or more focus time. The metrics become the neutral "truth" that allows the team to advocate for the environment they need to succeed.

The Leader's Role: Protect the Signal

When flow velocity drops, the leadership task is to understand what changed in the system and remove the obstacle behind it.

When teams trust metrics as a source of illumination rather than judgment, they stop hiding problems. They let the dashboard tell the truth because that is how the system improves.

Why it matters: Trusted metrics surface constraints early and guide targeted investment. Weaponized metrics distort behavior, hide reality, and turn leadership into surprise management.

Reality Check – When Output Races Past Outcomes

Atlassian's *State of Developer Experience 2025* reveals that, while AI usage has increased, half of developers still lose 10 or more hours each week to friction, and 63 percent report that leaders don't understand their pain points.[5] AI saves time in *Create*, but if *Ideate*, *Release*, and *Operate* remain chaotic, those gains are lost. The fix isn't to turn up the AI; it's to turn down the friction.

Shift 3 – Platform Thinking is a Leadership Responsibility

A good platform provides paved paths, sane defaults, and a reliable floor for security, compliance, and cost. Done well, it makes the preferred path easier to follow and reduces the need for teams to reconstruct the same decisions over and over.

A platform is not just a technical artifact; it requires product management, service promises, and executive sponsorship.

Platform as Product – Aligning DevOps Culture with Scalability

The phrase "you build it, you run it" has long been a DevOps mantra. In practice, it means teams deploy, monitor, and maintain their services. However, as complexity increases, this expectation can overwhelm product teams.

I have seen teams within an organization build entirely divergent infrastructure stacks, resulting in increased costs, fragmentation, and support chaos.

Turning platform responsibility into a product reclaims balance. Developers retain the DevOps mindset and accountability for quality, but rely on the platform to abstract away repetitive patterns. Platform-as-a-product becomes the bridge between the DevOps ideal of complete ownership and the reality of scalable, efficient organizations.

Case in Point – Spotify's *Backstage* Model

When *Spotify* developed *Backstage*, the platform created a centralized software catalog that made ownership, metadata, and service relationships easier to see across the engineering ecosystem.[6] That matters because platform thinking works best when teams can discover services, understand dependencies, and identify responsibility without hunting through scattered systems. The lesson is straightforward: when leaders treat the platform as a product, standardization can support autonomy rather than suffocate it.

When leaders neglect platform investment, autonomy collapses into chaos. Teams reinvent the wheel, security gaps multiply, and cloud costs spiral out of control.

During that period, the technology division treated platform infrastructure as a product, with a backlog and internal service expectations. Guardrails were firm where risk required them, and innovation was encouraged where possible. Paved paths, high automation, and clear boundaries increased delivery confidence in day-to-day work.

Shift 4 – Team Structures That Endure and Adapt

Long-lived, autonomous teams remain the cornerstone of healthy engineering systems. Static org charts may age, but stable teams with a clear product or value stream grow more capable over time. *Team Topologies* provided a useful language for thinking about stream-aligned, enabling, platform, and complicated-subsystem teams.[7]

The lesson is not to keep moving people, but to let durable teams evolve their boundaries. I did not favor dynamic reteaming by default. In my experience, strength came from continuity: from teams that stayed together long enough to master their domain, build trust, and refine flow.

The Impact of AI on Roles

AI is flattening some role boundaries. With *GitHub Copilot* and *Claude Code*, engineers can tackle adjacent tasks they once had to wait on, such as drafting test suites, scaffolding analytics, and mocking UI flows. Specialists still matter, but the baseline rises, allowing smaller squads to cover more ground.

Leadership failure here has two extremes: refusing to be flexible until structures break, or chasing fluidity so aggressively that careers and teams lose stability.

The future belongs to leaders who balance both: preserving long-term alignment, allowing short-term teaming where needed, and coaching people through transitions.

Shift 5 – Leadership as a Learning System

Psychological safety is not a perk. It is the precondition for performance. Teams learn faster when it is safe to surface risks, ask for help, and share half-formed ideas. Growth-mindset cultures scale that learning.

The half-life of decisions is shorter. With AI and platforms continually changing what is possible, leaders must be steadfast in their aim and flexible in their approach. Learning leadership shows up in habits: leaders admit what they do not know, run small experiments, and hire teachers, not only stars.

When leaders fail in this regard, the cost is high. Without safety, people stop speaking up. Without humility, experiments die. What should have been leverage becomes waste.

Rethinking Metrics – From Velocity to Value

Traditional metrics, such as story points and ticket counts, convey little about customer value and often create perverse incentives. They encourage leaders to push for motion rather than meaning. Future-fit leaders move away from velocity as the north star. Instead, they combine three lenses: Flow, DevEx, and Business Outcomes (Realization).

These three lenses mirror the core signals of a healthy value stream. When AI is applied effectively, its impact should be visible in reduced flow time, increased efficiency, reduced waste, and enhanced stability. The goal is not just speed, but business resilience.

Why it matters: Resilience is margin protection. The leaders who win will be the ones who can continuously prove how flow, quality, and outcomes move together, and intervene the moment they diverge.

Why It Matters to the Business: Multiple industry benchmark studies point in the same direction: organizations that improve delivery performance, reduce friction, strengthen platform and value stream practices, and connect delivery signals to business priorities are better positioned to improve outcomes over time.[8,9,10,11]

Treat the three lenses as a system. If flow time drops but defect escape rates climb, you are buying speed at the cost of instability. If DevEx improves but outcomes don't change, you have improved the experience without improving the value stream.

Leaders should establish concrete business metrics—adoption, retention, margin, incident impact—so that every platform or AI investment is visible in business terms, not just activity counts.

Building Leaders for the Next Decade

The next generation of engineering leaders will not be promoted solely on the basis of technical expertise. They will be selected based on their ability to think systemically, demonstrate learning agility, possess product acumen, and build trust across functions. They will know how to frame problems, not just solve them.

Classic management skills transfer cleanly to AI-assisted work. Set context, sequence work, match strengths to tasks, and give precise feedback. Some "resources" are now tools, but the leadership move is the same: diagnose with data and treat with design.

PUTTING IT INTO PRACTICE

Leading in this new era means shaping conditions, not commanding motion. To act on this, leaders can begin with five moves that turn systems thinking into daily practice:

1. *Read your system before you change it.* Spend time with one value stream map each quarter. Ask: Where does work wait? Where does efficiency erode? The goal isn't another ceremony, but visibility.
2. *Protect long-lived teams.* Audit how often teams are disrupted by external requests or reorganization. Build mechanisms that preserve continuity while improving collaboration.
3. *Use AI across the stream.* Review one workflow in each stage—*Ideate*, *Create*, *Release*, *Operate*—to see where AI can remove friction or waste. Ask: Does this shorten idea-to-value or merely automate local effort?
4. *Strengthen your platform partnership.* Meet regularly with platform and product leaders to review paved paths, developer experience data, and guardrails. Eliminate duplicate solutions.
5. *Link measurement to business conversations.* Tie flow and DevEx metrics directly to outcomes like adoption, margin, and retention. When these measures move together, the system is learning. When they diverge, intervene.

Key Takeaways

- *Design the system; do not manage the motion.* Shift leadership energy from supervising activity to shaping flow, outcomes, and safety.
- *Adopt AI as a baseline practice.* Treat AI as an amplifier with guardrails. Measure impact where value appears: flow, quality, and outcomes.
- *Invest in platform as a product.* Reduce cognitive load and standardize the floor. Keep guardrails firm and publish service promises.
- *Use a multi-lens model.* Combine Flow, DevEx, and business outcomes to avoid misleading single-metric stories.
- *Grow learning leaders.* Psychological safety and growth-mindset cultures are performance levers, not perks.

The work of tomorrow's leaders will depend on smarter orchestration.

If Chapter 10 is about leadership adapting to new systems, Chapter 11 is about the technology itself evolving into a new kind of collaborator. AI has become a cognitive amplifier that can reshape how teams think, create, and deliver.

The next chapter examines how organizations operationalize AI responsibly across the value stream, embedding it into ideation, creation, release, and operations. It focuses on the guardrails, measurement, and governance required to turn AI from a novelty into a durable operating advantage.

11 AI IN TECHNOLOGY: THE ORCHESTRATION EDGE

"Artificial intelligence is not a substitute for human intelligence; it is a tool to amplify human creativity and ingenuity." —Fei-Fei Li[1]

Quick Note on Recency

AI tooling evolves faster than most technology cycles. The specific model names, vendors, interfaces, and features will change. The point of this chapter is not to memorialize today's toolset, but to make the patterns durable: literacy, guardrails, measurement, operating model integration, and leadership discipline.

This chapter is about AI in the enterprise, where the test is no longer whether someone can build a local application or small demo with AI, but whether the organization can operate AI-enabled systems safely at scale without compromising quality, security, or trust.

This was the hardest chapter in the book to finalize, and one of the most exciting to write, because the ground kept moving under it. It was also the last chapter I wrote. I started it months before publication, and I rewrote parts of it through the final edit.

That volatility is not just a writing inconvenience. It is a leadership lesson. When the environment changes this quickly, the advantage is not perfect prediction. It is building option-rich systems: architectures, practices, and decision models that keep choices reversible, keep experiments cheap, and keep learning continuous.

The surface will keep moving. The deeper questions will not. How teams build trust, improve flow, design guardrails, govern risk, and connect technology work to business outcomes will matter long after better ones replace today's tools. The leadership job is to make speed safe, make outcomes measurable, and keep accountability human as the tools accelerate.

AI is a multiplier. In strong systems, it accelerates value, learning, and delivery. In weak systems, it accelerates defects, risk, and fragility.

THE ROLE SHIFT: DEVELOPER TO ENGINEER OF AUTONOMY

A second recency note matters even more than vendor churn: the role of the developer is shifting in real time. As generative tools become more agentic, the most valuable work moves upstream—from producing code to engineering the system that produces code safely. The keyboard is no longer the bottleneck. The bottleneck is whether your delivery system has the tests, telemetry, and guardrails to let automation accelerate flow without compromising quality, security, or trust.

To truly leverage what AI can do with code, enterprises have to get exceptionally good at the delivery system around it.

Leadership implication: When AI becomes a teammate, you inherit the responsibilities you would have with any powerful new hire. You have to onboard it with context, constrain it with standards, and hold it accountable through verification.

AI Literacy Before AI-First

When corporate leadership emphasized an AI-forward direction, the key challenge was that enthusiasm often outpaced definition.

We responded by focusing on durable principles such as literacy, guardrails, measurement, and workflow integration. This built upon the playbook we had already begun developing in early 2023, when we introduced *GitHub Copilot* and *ChatGPT* into developer workflows to assist thinking, not replace engineers.

We knew that if we waited for a formal enterprise rollout, fear would fill the vacuum. We saw two distinct anxieties emerge: the senior engineer, terrified that AI would devalue their decades of expertise, and the junior engineer, afraid to ask "dumb questions" about a tool everyone claimed to understand.

During 2025, as AI hype intensified, I heard many technology leaders declare "AI-First" initiatives as they tried to bring their organizations and divisions on board, often before the meaning of that shift was fully defined.

In practice, "AI-First" often sounded like a mandate to adopt the tools quickly and figure out the operating model later. The ambition made sense. The sequence was the problem.

We launched a specific mission to bridge that gap: AI Literacy by 2026. Lasting AI adoption begins with literacy, experience, and shared learning.

In my division, literacy did not mean blind adoption or success measured by prompt counts or AI-generated lines of code. It meant building enough understanding across the team to use AI in daily work, separate hype from reality, spot hallucinations quickly, and keep human judgment in charge. It was a way to make adoption safer, smarter, and more useful before scale turned enthusiasm into risk.

Leading by Learning (Vibe Coding)

We documented our own stumbles and published lightweight guidance on prompting, context, and hallucination handling so the team could see leadership learning in public. That visibility made experimentation safer.

When we first introduced AI into the Integrated Development Environment (IDE) in 2023, the productivity gains were real, but so were the risks: quality drift, constant review, and the feeling of supervising an intern who typed at lightspeed. That changed quickly. In a recent hands-on build, modern models and agents, used inside a disciplined workflow of constraints, iteration, tests, and verification, helped me deliver a meaningful full-stack slice far faster than I would have expected even a year earlier.

With modern models and agents, the job changed. Engineering still mattered just as much, but the center of gravity moved from typing to orchestration: intent, guardrails, and validation.

The Community Engine

We created a dedicated peer-learning space for experimentation and support, not as a policy broadcast channel, but as a place where people could share failures, surface risks, and learn in public. When broader governance was formalized, the team was not starting cold.

People were already more literate, less fearful, and better prepared to scale their use responsibly. That momentum gave us a shared language and let the team pivot quickly when the enterprise strategy formalized.

By then, team members understood that successful AI use depended on context, rules, and review.

From Literacy to Strategy

This experience shaped the chapter's core philosophy: adoption begins with literacy, not just access. At this stage, I am far more interested in literacy than slogans like 'AI-First.' The teams that get the best results are not the ones with the boldest mandate, but the ones that build experimentation habits around real work, open learning, problem-solving, and verification.

Before a broader enterprise strategy was fully formalized, we documented an interim adoption mission to orient the team. It was not a tool rollout. It was a way to evaluate AI decisions through the same lenses that guide the work: Flow, Realization, and Quality.

The principle was simple: AI adoption should strengthen, not bypass, the systems that already define excellence.

Why it matters: An AI strategy without the right measurement becomes theater. This mission keeps investment tied to outcomes. AI increases throughput and learning without turning quality, security, or trust into collateral damage.

SOONER SAFER AI

If AI is a capacity multiplier, governance becomes the constraint that determines whether that capacity becomes value or risk.

One of the most persistent myths about security and compliance is that they slow innovation. That myth survives because many organizations apply controls too late and too manually. They treat safety as a gate at the end instead of a system property designed from the start.

AI makes that approach unsustainable.

If guardrails arrive after experimentation, leaders are forced into a false trade: speed or safety. Teams learn the wrong lesson: move fast until someone tells you to stop. Shadow usage grows. Trust erodes. And when the shutdown comes, it feels punitive, not protective.

Sooner Safer AI flips the sequence. It is the discipline of designing safety into the system early so that teams can move quickly within clear constraints.

Context matters. Sooner Safer AI does not mean every organization needs the same controls at the same depth. The right guardrails depend on context: the type of work being done, the sensitivity of the data, the level of regulatory exposure, and how directly AI is being used in production delivery or customer-facing decisions.

Government systems, financial software, and a general e-commerce site do not carry the same risk profile, and they should not be governed as if they do.

A team using AI to summarize notes or draft internal documentation does not face the same consequences as a team using agents to modify code, handle sensitive customer data, or influence operational decisions.

The principle stays the same, but the depth of governance should match the risk and the potential consequences of failure.

As AI moves from assistance to orchestration, governance must mature alongside it. Clearer boundaries, better visibility, and faster intervention become essential.

Agents need a defined scope, known permissions, strong defaults, and observable behavior. When they hit a guardrail, the system should not just keep retrying in the dark. Someone should be able to see what happened, understand the context, and decide what comes next. The more autonomy leaders allow, the more discipline the surrounding system must provide.

The temptation in an AI wave is to lead with tools: grant access, encourage experimentation, and figure out governance later. That sequence almost always backfires. When rules arrive after behavior, they create fear. When rules arrive before adoption, they create clear boundaries for experimentation.

Sooner Safer AI starts with a simple leadership move: define what "safe" looks like, and make the safe path the easiest path. This is responsible autonomy. It is the same pattern that enables fast flow in modern delivery systems: freedom inside constraints.

In practical terms, guardrails should answer questions teams should not have to guess:

- What data is safe to use, and what data is off limits?
- What types of work are acceptable for AI assistance, and what work requires additional review?
- What must remain a human decision, even when AI can suggest an answer?
- What evidence must exist for a change to be trusted?

Traditional governance relies on documents, training, and approvals. That is too slow for an environment where AI can generate artifacts instantly, and agents can take actions across systems. A more durable approach is to move from policy as a document to policy as a default. This is policy-as-code and controls-as-code thinking applied to AI adoption.

When guardrails are encoded into workflows, the safe path becomes the easiest path. Access boundaries are enforced automatically. Sensitive data is protected by default. Audit trails and escalation points are built into the system rather than reconstructed after the fact. This is how compliance becomes the paved road instead of the speed bump.

That shift also changes how leaders should think about compliance itself. In many regulated environments, compliance has historically meant documentation. AI exposes the limits of that model. Documentation can be written after the fact.

Evidence is harder to fake. A stronger model is evidence-based compliance: clear proof of which data were used, which controls were applied, what remained a human decision, and which tests, reviews, and rollback mechanisms supported the change.

This is visibility, not bureaucracy. It is the same discipline modern delivery systems use when they move from "trust us" to "here is the proof." It also aligns with the operating lenses throughout this book: Flow makes work visible, Quality makes work safe and resilient, and Realization makes value measurable.

When security is built as evidence, it stops being an argument and becomes part of the same feedback system as delivery itself.

When those boundaries are unclear, teams invent their own. That is how risk spreads invisibly.

CUTTING THROUGH THE NOISE

AI has arrived in software engineering with enormous hype. The headlines are familiar: developers will soon be obsolete, prompts will generate enterprise-grade systems, and operations will run themselves.

The reality is both more modest and more transformative. Today, AI behaves more like another teammate in the system than a substitute for engineers. It accelerates boilerplate, scaffolding, debugging, documentation, and other non-differentiating work.

What it does not replace is the discipline, architecture, and culture required to deliver secure, resilient, and scalable software.

For leaders, the question is no longer whether AI will change software delivery. It already has. The question is whether you will frame adoption around outcomes, such as flow, value, and developer experience, or fall into legacy habits of cost-cutting and vanity metrics.

I recently saw a comment aimed at engineering leaders: 'Only cost centers brag about cost savings.' It is a useful caution. When organizations treat AI mainly as a reason to cut investment, they turn potential into a constraint. More mature organizations treat AI as a lever for acceleration and reinvest efficiency gains into innovation, customer outcomes, and system health.

AI can also tempt leaders back into an old management reflex: mistaking visible output for meaningful progress. Dashboards that count prompts, track AI usage, or compare AI-assisted pull request volume may look precise, but they still measure activity at the edge of the system.

They do not tell leaders whether value reached customers faster, whether quality held, or whether the delivery system became healthier. A team can generate more code and open more pull requests while testing capacity, release cadence, and operational readiness remain the real constraints. In that situation, the organization has not improved flow. It has simply pushed more change into a system that may not be ready to absorb it.

As discussed in Chapter 4, Goodhart's Law applies here as well: once a measure becomes a target, people start optimizing the score instead of the purpose.

Reward prompt volume, and usage will rise whether it helps or not. Reward percentages of AI-authored code, and some teams will optimize for visible output rather than sound engineering.

The leadership task is not to reward AI activity for its own sake. It is to determine whether AI is reducing friction, protecting quality, and improving realized outcomes across the value stream.

AI makes engineering more capable and more demanding at the same time, and that divergence is already showing up across the industry.

By 2025, the leadership discipline that mattered most was no longer usage alone, but measurement and feedback. It was no longer enough to ask how many teams were using AI. Leaders had to ask whether AI was improving flow, quality, and customer value. Usage rates measure activity. Feedback loops measure impact.

In practice, that maturity shows up when leaders stop treating AI as a narrow productivity gain and start watching how it changes the balance of the whole value stream.

Software delivery can break in two directions: it can get starved or jammed.

On one end, if the backlog can't produce enough credible, well-shaped work, teams get starved. The fuzzy front end (discovery, prioritization, clarification, sequencing) becomes the constraint. AI can help generate options and accelerate analysis, but if it doesn't materially improve idea-to-backlog readiness, implementation speed just creates idle time, thrash, or lower-value work pulled in to stay busy.

On the other end, when implementation accelerates, the system can still jam downstream. More changes hit review, test, validation, security, and deployment. Pipelines get noisier, queues get longer, and feedback slows. The pain doesn't always show up in the dashboards, but teams feel it immediately.

The takeaway is simple: speed in one stage doesn't equal flow. The next real gains come from balancing the full value stream by helping the backlog generate ready work at the pace teams can build, and helping CI/CD, testing, and release systems scale to absorb what gets built safely.

That's why the approach focused less on tool excitement and more on habits: literacy, guardrails, and workflows that improve the system, not just the typing.

VIBE CODING: BUILDING EVERYDAY AI FLUENCY

During that period, the division I led approached adoption with a simple philosophy: Team members treated AI like a teammate, with real expectations for context, discipline, and review. We adopted *Vibe Coding*, an industry term that captures the fluid, collaborative nature of working with AI tools.

Principles of Vibe Coding:

- Learn by doing: Apply AI on real bugs, features, and documentation.
- Provide structure: Product Requirements Documents (PRDs), tasks, and rules give AI the necessary context.
- Stay accountable: AI drafts; humans decide. Quality and security remain with us.
- Share the journey: Document successes and failures to accelerate learning and growth.
- Support the hesitant: Adoption must be beginner-friendly; confidence comes from practice, not pressure.

Why it matters: These principles prevent 'AI speed' from becoming hidden rework later. They protect flow and quality by keeping accountability human while letting automation remove the grind. In that sense, Vibe Coding succeeds when leadership creates safe conditions for experimentation, builds confidence, and embeds AI into the culture without hype or fear.

A Tale of Two Developers

```
Developer (open PR w/ template)
   ↓
AI Gate Check (Copilot / Claude)
   ↓
Fix gaps (tests, evidence, observability)
   ↓
CI/CD Check (GitHub Action)
   ↓
Human Reviewer (design, correctness, risk)
   ↓
Merge to trunk or release branch
```

Figure 11.1: *The Vibe Coding Workflow.* The modern development loop integrates AI gate checks and automated gap-filling before human review, shifting the developer from syntax writer to logic orchestrator.

To understand the shift, contrast the workflow of a "Legacy Developer" with a "Vibe Coder."

The Legacy Workflow (The "Before"):

A developer picks up a ticket for a new API endpoint and spends much of the effort on syntax, boilerplate, searching for answers, and working through avoidable friction. Testing happens later than it should, feedback arrives late, and quality issues surface downstream. Result: high effort, low flow, cognitive fatigue.

The Vibe Coding Workflow (The "Now"):

The developer picks up the same ticket but starts with intent and context inside an AI coding assistant. The tool quickly scaffolds the endpoint, boilerplate, and an initial test suite. The developer shifts into reviewer mode, validating logic, security, and quality rather than hand-authoring every line. Tests run earlier, defects surface sooner, and the pull request moves with more complete context. Result: high velocity, high flow, cognitive focus on quality.

This shift turns the developer from a "writer of syntax" into an "orchestrator of logic."

And when the individual workflow changes, so do team dynamics.

AI changes team dynamics by changing how work is produced, reviewed, and distributed. When agents take on boilerplate, first-pass documentation, and repetitive test scaffolding, teams get more time for the work that still requires humans: clarifying intent, debating tradeoffs, challenging assumptions, and aligning on the why.

It also creates friction during adoption. People move at different speeds, and reviews can become emotionally charged when one person sees a useful draft and another sees a dropping standard.

The answer is not to ban the tool or worship it, but to make quality explicit and shared through tests, security checks, readability, and clear intent. Accountability remains human, and AI fluency becomes a team capability that must be learned together.

AI also blurs role boundaries. Done well, that can reduce handoffs, increase swarming, and make teams more resilient. Done poorly, it creates a false sense of coverage.

The discipline is to encourage cross-functional contribution while keeping ownership of design integrity, risk decisions, and production readiness clear. Roles do not disappear. They evolve.

As agents begin to take on role-shaped work across the team, the advantage does not come from pretending the roles no longer matter. It comes from helping each discipline understand what AI can accelerate, what still requires human judgment, and how collaboration changes when more of the execution is assisted or orchestrated.

The Talent Divergence: Acceleration vs. Illusion

A pattern is emerging across the industry, one that every hiring manager and leader must recognize: AI is creating a *divergence* in talent.

Senior engineers and those with strong architectural instincts are achieving massive productivity gains. They can guide AI systems, correct drift, and explain clean architecture to the tool. For them, the AI is a force multiplier.

However, for engineers who lack a strong foundation, AI creates a dangerous archetype: *The Glass Cannon Engineer.* These engineers produce impressive-looking code at incredible speed. Still, because they lack the deep understanding to spot hallucinations or security flaws, their work shatters under the pressure of production debugging. They produce legacy code faster, creating an *illusion of velocity* that is actually the rapid accumulation of debt.

This creates a *"Dual Bar Hiring"* for modern engineering talent. We cannot choose between "Fundamentals" and "AI Fluency"; we must demand both.

- *Fundamentals:* Can they reason through a system design without the tool?
- *Fluency:* Can they direct the tool to execute that design with precision?

AI raises the floor for output, but it raises the ceiling for the judgment required to manage it.

Why it matters: This is a governance and hiring issue, not just a tooling shift. If leaders don't protect fundamentals, AI will manufacture the appearance of progress while accumulating risk and long-term cost.

Leadership Tool: The Dual-Bar Hiring Rubric

AI raises the ceiling of what strong engineers can do. It also masks weak fundamentals. That is why the hiring bar must evaluate both capability and amplification.

Use this rubric to assess modern talent without rewarding "AI-shaped theater."

Fundamentals (Non-Negotiable)

- Systems thinking: Can they reason about trade-offs, boundaries, dependencies, and failure modes?
- Problem decomposition: Can they break ambiguity into coherent, testable steps?
- Debugging discipline: Can they form hypotheses, isolate variables, and explain why something failed?
- Design reasoning: Can they explain why a design works, not just what they would build?
- Learning posture: Can they name what they don't know and describe how they would find out?

AI Fluency (Capability Multiplier)

- Intentful prompting: Can they provide context, constraints, and desired outcomes rather than fishing for answers?
- Workflow integration: Can they explain where AI fits in real delivery work (design, implementation, testing, review, documentation)?
- Verification discipline: Do they validate AI outputs through tests, reasoning, or alternative sources?
- Judgment under speed: Can they move faster without skipping thinking?
- Boundary awareness: Do they know when not to use AI (sensitive data, unclear requirements, high-risk changes)?

Integrity Checks (Trust Signals)

- Attribution honesty: Do they clearly distinguish their reasoning from AI-generated suggestions?
- Error awareness: Can they recognize when AI is likely wrong and say so early?
- Safety instincts: Do they respect privacy, security, and compliance boundaries without being prompted?
- Ownership mindset: Do they take responsibility for outcomes even when AI is involved?
- Explainability: Can they explain decisions to a peer reviewer and to a non-technical stakeholder?

How to Use This Rubric

- Do not average the bars. Fundamentals are table stakes.
- Interview for thinking, not typing. Ask candidates to narrate decisions and verification steps.
- Probe AI use explicitly. Ask when they would not use AI, and why.
- Hire for amplification, not substitution. AI should increase judgment and learning speed, not bypass them.

This is how leaders avoid "AI-fluent, fundamentally lost" teams while still capturing the compounding upside of AI-enabled delivery.

FROM GENERATIVE TO AGENTIC

The labels in this space will continue to shift. What is called generative, agentic, assistive, autonomous, or orchestration-driven today may be renamed, recombined, or absorbed into the next wave of tools tomorrow. The important point is not the label. It is the direction of travel.

As teams built fluency, the technology advanced just as quickly. The generative wave of 2023 and 2024, centered on writing code and text, is evolving into agentic systems that do more than assist developers. They increasingly orchestrate the workflow itself, planning, taking actions, and coordinating work across tools. We are moving from one assistant to coordinated agent teams, and the capability jump is real.

However, leadership must see the other side of that curve. These systems can burn a lot of tokens and money, especially when agents run continuously. That makes cost management a real operational discipline, not a finance afterthought. The right comparison is cloud economics. Just because you can scale does not mean you should do it unmetered.

Long-running agents are impressive, but they can drift, loop, or accumulate subtle mistakes. Enterprises will depend on them at scale only when the harnesses mature: checkpoints, evaluation gates, kill switches, and strong evidence from tests and telemetry.

This shift will not only change how we build, but how we shape teams. In some cases, the most effective unit may become smaller and more direct: the person with the problem and the person who can fix it, working together with AI to compress the distance between intent and implementation.

The names will change. The pattern is stable: agency increases capability, but it also increases the need for guardrails, measurement, and clear accountability.

Looping is the new shape of work. More work now happens in feedback loops rather than linear steps: intent is defined, the system proposes changes, automated checks return feedback, and the loop repeats until the evidence says 'done.' That is why guardrails and fast feedback are no longer supporting practices. They are the steering mechanism.

The organizations that win will not be the ones with the loudest AI mandate, but the ones that turn feedback into a disciplined operating rhythm so autonomy can scale without heroics.

For my division, that transition was still unfolding. We were beginning to apply and evaluate agentic patterns in safe, repeatable work such as dependency upgrades, test generation, risk analysis, and modernization. Engineers curate results, add domain judgment, and use Flow Metrics to measure impact.

AI and the Value Stream

AI will not fix a poorly designed operating model; it will amplify it. When ownership is unclear and decision-making is bottlenecked, AI increases activity without improving outcomes. When teams have bounded autonomy and clear interfaces, AI accelerates learning and delivery. That finding is not unique to us. The 2025 DORA analysis reached the same conclusion: AI does not change the system; it magnifies it.[3]

Enterprise AI is a system conversation across the full value stream.

There is a familiar warning here for leaders who have lived through other transformation waves. Teams can adopt new tools while the surrounding system stays largely unchanged: tangled architecture, unclear boundaries, heavy governance, and local incentives.

When that happens, the method gets blamed for conditions it never had the power to fix. That is why what matters next is the system around the work, not the assistant in isolation.

Even if AI makes rewrites cheaper, architecture still matters because rewriting is a worst-case option, not a default operating model. As output gets cheaper, coordination gets more expensive.

The real bottleneck becomes the engineering system itself: tests, telemetry, review discipline, backlog clarity, CI/CD, platform design, release confidence, security, and operational readiness. When those conditions are weak, AI speeds up the system, which means unresolved constraints surface faster.

The Agentic Prerequisite: DevEx

Many organizations still treat platform work as a collection of tools. That was limiting before AI. It is a much bigger liability now.

When teams can generate more code, automate more steps, and experiment more quickly, they need a platform that feels coherent, usable, and intentionally designed.

The platform is not just infrastructure. It is the delivery experience: paved roads, safe defaults, reusable workflows, and feedback loops that help teams move faster without creating chaos. AI will magnify whether that experience is clear and supportive or fragmented and frustrating.

That is why developer experience has become more than a productivity concern. It is now part of AI readiness. Documentation, modularity, clean requirements, and fast build pipelines are no longer just good discipline; they are essential. They shape whether people and agents can work effectively inside the same system.

One subtle constraint appears quickly as teams move from assistive prompts to agentic workflows: context is capacity. Agents perform well only when the working set remains coherent, the scope is bounded, and verification is fast.

Larger context windows do not eliminate that problem; they often change its shape. Retrieval is still imperfect, and too much context can create drift rather than clarity. The best teams manage context with the same discipline they bring to engineering.

One practical pattern is treating your AI like a new teammate that needs onboarding. Do not expect it to infer your design sensibilities from an entire codebase every time. Give it the rules of the road. Put your architectural preferences, coding standards, testing expectations, and security constraints in short, explicit documents that the model can reliably consume. Humans need onboarding to be effective. AIs do too.

Agentic development changes the calculus. The prerequisites for AI success look like the conditions required for human success.

Humans need context; so do LLMs. If requirements are vague, an agent will hallucinate just as quickly as a junior engineer will guess wrong.

Humans need structure; so do LLMs. If the codebase is a tangle of dependencies, agents cannot reason through changes effectively.

Humans need feedback; so do LLMs. Without fast, automated feedback loops, agents cannot self-correct, at least not reliably yet.

What is good for the human developer is essential for the synthetic one. Investments in DevEx, clean APIs, updated documentation, and fast build pipelines are no longer just about developer happiness. They are about organizational readability.

AI offers enormous potential, but it introduces new risks if layered over a broken foundation. To succeed, leadership must ensure its investments do not just buy new tools, but actively improve the environment in which those tools operate.

The goal is to create conditions that help teams deliver high-quality, secure, and resilient value at the speed many businesses demand.

REALITY CHECK:
Atlassian's *2025 State of Developer Experience* research underscored the same paradox: developers reported double-digit time savings with AI, yet lost a similar amount of time to organizational friction.[4] Measurement and feedback loops, not tooling alone, determine net gain.

Putting Agents Inside the Flow

One of the more promising frontiers for AI adoption is its integration into Flow Engineering practices.

Today, our teams use data and mapping to identify constraints. In the future, we envision pairing those methods with intelligent agents that operate *within* the flow rather than simply *on top of it.*

We are moving toward a model where AI agents assist across the entire value stream:

- *Ideate:* Agents synthesize research, draft business cases, and model risks before a project begins.
- *Create:* Agents handle scaffolding, boilerplate generation, and real-time peer review.
- *Release:* Agents automate compliance checks, dependency analysis, and deployment choreography.
- *Operate:* Agents detect incident patterns, suggest diagnostics, and feed learnings back into the backlog.

The idea is simple but powerful: AI agents act as collaborators that make invisible work visible—spotting dependencies, surfacing insights, or simulating potential changes before teams commit effort. Rather than replacing analysis, they offer data-driven hypotheses that help teams move faster.

Systems that think with you eliminate decision drafts, reduce rework, and unlock predictability.

The Agent in the Operations Center

Let's look at exactly how this changes the *Operate* phase—often the most stressful part of the lifecycle.

In a traditional setup, when a P1 incident alert fires at 2:00 AM, the on-call engineer wakes up groggy. They spend the first 20 minutes logging into dashboards, scrolling through logs, and trying to correlate the error with recent changes. "Was this the database migration? Or the new front-end deploy?"

In an agent-augmented workflow, the AI is the first responder.

1. *Detection & Triage:* The agent detects the anomaly. Instead of just paging the human, it immediately pulls the relevant logs and correlates the timestamp with the last three deployments.
2. *Hypothesis Generation:* It analyzes the stack trace and identifies a likely cause (e.g., "High latency correlates with the recent cache configuration change in Service B").
3. Contextual Hand-off: When the engineer accepts the page, they don't see a blank screen. They see a summary: "Incident detected in Service B. Likely cause: Cache miss spike. Recommended Action: Rollback Release v2.4 or increase TTL."
4. *Remediation & Loop Back:* The engineer validates the hypothesis and authorizes the rollback. The agent executes the command and then drafts the initial postmortem, tagging the specific commit that caused the drift.

The human remains the decision-maker, but the agent removes the "fog of war." This is how we compress Mean Time to Restore (MTTR) and feed learning back into the system. Software does not stop at deployment. If AI-driven efficiency reduces operational depth to the point where only a few people can safely recover the system, the result is concentrated operational risk.

Why it matters: Operational speed is profit protection. Faster diagnosis and recovery reduce downtime costs, protect customer trust, and return engineering capacity to innovation rather than incident churn

Many Value Stream Management (VSM) and Software Engineering Intelligence (SEI) platforms now include embedded AI capabilities, pointing toward a shift from dashboards to more strategic intelligence.

We are not fully there yet, but the direction is clear. As data quality and guardrails mature, AI and agents can become an extension of Flow Engineering itself, helping analyze patterns and shorten feedback loops faster than manual analysis alone.

Agent demos are seductive, but production reliability is earned through discipline, not spectacle. When systems can rapidly generate significant changes, manual review cannot be the only safety mechanism.

Verification has to be built into the system through test evidence, progressive delivery, and rollback-by-default release practices. Humans remain accountable, but the system carries a greater share of the assurance burden.

Case Study: Modernization at Scale

As AI adoption matures, one of the most practical opportunities lies in modernizing legacy systems, where automation and human oversight must work hand in hand.

In complex fintech environments, engineering teams have combined deterministic transformation libraries, such as *OpenRewrite*, with AI agents that orchestrate code upgrades. These agents select “recipes” (pre-defined, auditable transformation rules) and apply them across hundreds of repositories under strict validation. This hybrid model strikes a balance between automation and safety.[5]

At enterprise scale, Morgan Stanley’s internal *DevGen.AI* platform demonstrates how targeted AI accelerates transformation. The system translates legacy code (such as COBOL) into standardized specifications. *Business Insider* reported that *DevGen.AI* saved more than 280,000 developer hours, showing how deeply AI can integrate into modernization when paired with rigorous governance.[6]

The lesson is not that every organization should rush to automate refactoring; instead, the combination of repeatable transformation logic, human validation, and measurable outcomes defines safe modernization at scale.

Lessons from Peers

The experience in my division aligns with a broader pattern across engineering organizations: the teams seeing the best results do not treat AI as a shortcut to productivity. They integrate it into the systems, culture, and measurement practices that already support strong delivery.

Across these organizations, the pattern holds: AI works best when it amplifies shared understanding, flow, and innovation across the value stream, not just productivity within delivery. Effective leadership in the AI era, therefore, requires clear priorities, experimentation, and accountability.

THE LEADERSHIP EQUATION

Leaders must enable funding, protect time for learning, and model curiosity by exploring AI themselves, not as experts, but as participants. In organizations where leaders engage directly in learning, testing, and review alongside their teams, adoption gains trust and momentum more naturally.

The most effective executive sponsorship does not come from mandates. It comes from leaders who use the tools themselves and then ask the obvious question: if this improves my work, why would it not improve my teams'?

The broader principle remains unchanged: leadership's role is to make outcomes clear, reduce wait states, and maintain transparency. As AI brings more visibility and faster feedback loops, the pace of learning increases. That change demands humility and adaptability from leaders, as well as the ability to listen to the system as it speaks back through data.

For leaders, the next move is not another AI slogan. It is an honest assessment of whether the system is ready to benefit from AI without creating more noise.

That means looking at delivery friction, platform coherence, data readiness, quality controls, security hygiene, and team clarity.

It means choosing a few meaningful use cases, measuring whether flow, confidence, and outcomes improve, and leading adoption in a way that builds judgment across the organization rather than just enthusiasm.

PUTTING IT INTO PRACTICE

Leading in this new era means shaping conditions, not commanding motion. To act on this, leaders can begin with eight moves that turn systems thinking into daily practice:

1. *Revisit and communicate your strategy often.* Treat AI as a living system. Build cycles to revisit assumptions and realign your message.
2. *Anchor AI in the existing strategy.* Integrate AI into *Flow*, *Realization*, and *Quality*. When AI ties to the same outcomes that define success, it becomes a natural extension of work.
3. *Apply AI across the value stream.* Think beyond engineering. The most significant leverage appears when AI assists every function from ideation, design, finance, and operations.
4. *Strengthen guardrails first.* Establish boundaries for quality and security. Define "responsible autonomy" and scale AI only as safeguards mature.
5. *Measure system impact, not activity.* Track metrics like impact and cost, but connect them to throughput, flow efficiency, and DevEx. The goal isn't to prove AI is used; it is to prove AI enhances flow.
6. *Measure toolchain risk.* Set cost ceilings, maintain continuity plans, and watch vendor dependency, because AI is becoming part of your production delivery system.

7. *Lead by example.* Experiment in your own work. Modeling exploration permits others to learn.
8. Make AI part of continuous improvement. Incorporate AI reflections into retrospectives. Ask: What friction did it reveal?

Key Takeaways

- *AI is an integral part of the delivery system.* Healthy systems accelerate with AI; weak ones amplify chaos.
- *AI is a dependency now.* Manage cost, continuity, and retrieval limits like you manage cloud spend and production risk.
- *Licenses don't create literacy.* Culture, education, and psychological safety do.
- *Autonomy demands rigor.* Tests and reviews must strengthen alongside capability.
- *Measure what matters.* Focus on flow, DevEx, quality, and value-stream compression—not just prompt counts.
- Leadership must model adoption. Credibility outlasts slogans.

The story of AI in technology is, at its core, the story of transformation itself. It is about how organizations learn to see their systems clearly, align around outcomes, and evolve faster than their constraints. AI is one inflection point in that journey, not the destination.

AI is moving fast. The leadership edge comes from the system: clear intent, strong constraints, and verification that turns output into evidence. Leaders who build learning systems will move faster. Leaders who chase tools will stay behind.

Across this book, we have explored what it means to build a modern technology organization: one grounded in flow, connected to business value, and guided by leaders who make priorities, decision rights, and accountability visible. We have moved from legacy thinking to systems thinking, from project outputs to measurable outcomes.

Every practice (Agile, Lean, DevOps, Value Stream Management, Flow Metrics, and now AI) has been a layer in a larger framework of purpose: to turn technology from a cost center into a strategic partner.

That is the real journey ahead. It starts with understanding your system, strengthening your flow, and building a culture that learns faster than the market changes.

Your journey starts here—not with new tools, but with new intent.

The real advantage will not come from tool access alone. It will come from building systems strong enough for AI to make them better.

12
YOUR JOURNEY STARTS HERE

"In today's era of volatility, there is no other way but to re-invent. The only sustainable advantage you can have over others is agility, that's it. Because nothing else is sustainable, everything else you create, somebody else will replicate." —Francisco D'Souza[1]

The Long Arc of Reinvention

Every system ultimately reflects its builders.

I still remember the hum of those first servers in the early days of the internet, the kind that sounded more like a box fan than a gateway to the future. We were writing code to keep a fledgling e-commerce site alive, racing against the technological capabilities of the time and my own inexperience. We weren't talking about "digital transformation" or "continuous delivery." We were trying to make the thing work before the next outage or peak customer season.

Back then, technology felt mechanical: build a feature, ship it, and hope it didn't crash. But those early days marked the beginning of a much longer journey, one that would carry me from small startups to global enterprises.

What changed most over those decades wasn't the technology itself; it was us. The way we lead, the way we measure progress, and the way we connect our work to value.

The Crucible of Transformation

I didn't learn the lessons in this book in a classroom; I learned them in the trenches at Parchment.

From what I recall, when I joined in early 2012, we were around 50 people with roughly $5 million in revenue. I started with a team of just eight engineers. We were operating in a world that many legacy enterprises would recognize today: a massive, monolithic platform, a "Water-Scrum-Fall" culture, and a release cadence defined by pain. We had two massive releases a year, a defect backlog numbering in the hundreds, and functional silos that passed responsibility back and forth like a hot potato.

Between 2014 and 2016, we decided to stop patching the old way of working and start architecting a new one. We strangled the monolith, architecting it into a cloud-based microservices architecture. But as Conway's Law predicts, we couldn't fix the code without fixing the teams.[2]

We moved from functional silos to fully self-managed, cross-functional teams aligned to domains and value streams. We shifted quality left, adopting stronger quality engineering and testing practices, and we made a hard rule: stop managing defects and start fixing them. We systematically drained that

800-defect backlog to near zero, treating new bugs as unfinished work rather than backlog items.

As I grew into the Head of Engineering role, we didn't just scale the code; we scaled the organization. We grew from a small engineering team to a global organization of well over a hundred engineers. The company itself expanded its headcount nearly tenfold. That growth wasn't purely organic; during my tenure, we also completed multiple acquisitions, where I led or contributed to technical due diligence.

Because we had built a scalable product operating model rather than a rigid hierarchy, we were able to digest those acquisitions and continue accelerating.

- *Speed:* We went from 2 releases a year to continuous deployment, with multiple production releases daily.
- *Quality:* We maintained the zero-defect mindset, achieved near-five-nines reliability, and stopped technical debt from accumulating as inventory.
- *Flow:* We abandoned Gitflow in favor of trunk-based development, reducing lead times from months to minutes. We moved from Scrum to a Kanban-based workflow tailored to our context and delivery constraints.
- *Platform:* Long before "Platform Engineering" was a buzzword, our Production Engineering team treated our internal developer platform as a product, providing a "paved road" that allowed developers to move fast without breaking things.

During that period, Parchment became the poster child for digital transformation. But on the inside, it was a grind. We struggled. We had to fight for every inch of that progress against the gravity of organizational inertia.

But the results of that struggle were undeniable. By the time of the sale, revenue had grown substantially from where it stood when I joined. The company ultimately achieved a publicly reported valuation of approximately $835 million at the time of acquisition.[4]

Technology was part of that success—but it wasn't the sole cause. No amount of DevOps maturity can replace a great sales team or true product–market fit. What *Profitable Engineering* provided was asset quality.

We didn't bring investors a perfect platform. We built a modern, competitive foundation by integrating acquired capabilities into a coherent, modern architecture supported by consistent teams, culture, and delivery practices. So when diligence happened, they didn't see a brittle legacy estate that required a ground-up rewrite; they saw a delivery engine with momentum, capable of absorbing growth and accelerating product execution.

And when Sales found the market, our practices ensured we could capture it. That's the difference between a cost center that constrains growth and a strategic partner that secures it.

That is the power of Profitable Engineering. It isn't just about shipping code faster; it's about building a system resilient enough to scale a business and proving its value to the market in hard currency.

The Culture of Space: Why Innovation Needs Room to Breathe

While structural changes to teams and architecture were critical, they were not enough. We realized that every great system ultimately depends on one thing—its people.

Processes, metrics, and technology matter, but they only reach their potential when the people using them feel trusted, empowered, and supported. At the heart of our transformation was a simple belief: teams outperform individuals. We stopped celebrating heroics and started celebrating collaboration. But you cannot simply command people to "be innovative." You have to create the space for it.

One of the most impactful ways we reinforced this was through our own version of Google's 20% time. Once a week, teams could dedicate time to exploration, innovation, or improvement projects outside of their day-to-day work. We didn't mandate what they worked on. Some used the time to refactor critical code paths that had been annoying them for months; others prototyped automation tools, and some used it simply to learn and share new technologies.

Over time, this practice became more than a policy; it became part of our DNA. It reminded everyone that innovation doesn't require permission; it needs space.

Crucially, this shifted how leadership listened. I learned that the best ideas didn't always come from senior meetings or strategic plans; they came from hallway conversations, retrospectives, or product demos where someone asked a simple question that reframed the problem entirely. The ideas generated from those Fridays often became production improvements or new features.

This culture of "space" was the hidden variable. It was the reason we could pivot so quickly when the next wave of disruption hit.

The Acceleration of Change

If Parchment was the crucible where I learned resilience, the explosion of generative AI in 2023 was the test of engineering's agility.

When generative AI emerged as a serious force in early 2023, we didn't wait for a corporate mandate or a vendor roadmap. We relied on the culture of continuous learning, the muscle memory built during our weekly innovation time, to move fast. We launched AI literacy initiatives across the R&D organization, and within months, we weren't just talking about AI; we were running experiments and rewriting our playbook in real time.

The contrast was stark. The transformation at Parchment took nearly a decade. The real AI pivot took less than six months.

This reinforced the most critical lesson of my career: *The cycle of reinvention is compressing.* What used to take years now takes quarters. Leaders no longer have the luxury of "waiting to see what settles." You are either learning faster than the market or already falling behind.

That was the pivot point. I realized that being a great engineering leader wasn't enough. I had to become a *Value Stream Executive.*

THE CONVERGENCE: FLOW, FUNDING, REALIZATION, AND AI

Throughout this book, four threads have emerged that leaders must engineer to move together: Flow (how work moves), Funding (how incentives shape behavior), Realization (how value is proven), and AI (how capability scales). Together, these form what I call the Value Stream Convergence Model.

In isolation, each of these is powerful. But your journey as a modern executive begins when you fuse them into a single operating model.

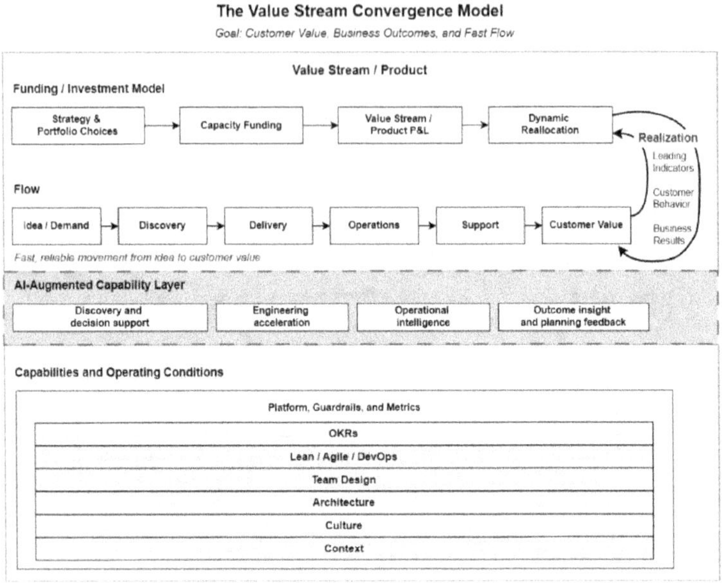

Figure 12.1: *The Value Stream Convergence Model.* True profitability emerges when we connect Strategy (Realization) with Operations (Flow). This model illustrates how foundational investments from Culture to Architecture support the model's high-level goals.

The lessons from Parchment taught me that Flow can survive imperfect funding models, but it becomes harder to protect when the funding mechanism rewards projects over products.

I've heard variations of a goal that reveals how deeply cost-center thinking is embedded: make R&D "cost neutral." When that phrase surfaces, it is always a wake-up call. When you're still operating inside annual budgets and CAPEX/OPEX mechanics while trying to run durable product teams, you can make it work. Still, you're relying on leadership discipline to keep the system from snapping back into project reflexes.

As discussed in Chapter 9, this is what a Value Stream P&L enables: investment stays with the product or value stream and the teams that own it, allowing leadership to govern with an economic lens rather than by delivery activity alone.

But Flow and Funding still are not enough if you never verify that the investment changed customer behavior or business results. That is what I mean by Realization: the discipline of closing the loop between delivery and measurable business impact. It is how technology becomes legible to the business and credible to the boardroom.

And you cannot maximize that value without AI. As we explored in Chapter 11, we are entering the age of talent divergence and agentic workflows. Leaders who treat AI as a coding assistant will see marginal gains. Leaders who treat AI as a systemic capacity multiplier, integrated into the value stream with guardrails, will see compounding returns.

In modern organizations, Flow, Funding, Realization, and AI must be engineered to move together. This is the convergence.

- *Flow* is how you deliver.
- *Funding* is how you align incentives.
- *Realization* is how you prove value.
- *AI* is how you scale capability.

These four forces converge naturally when the system is designed for it. They fragment when leadership defaults to legacy reflexes.

THE VALUE STREAM EXECUTIVE IN ACTION

So what does this convergence look like in the boardroom? This is the work of the Value Stream Executive: a leader who connects Flow, Funding, Realization, and AI into one business conversation. It requires a fundamental change in the dialogue between engineering and the business.

In the old world, the world of the cost center, the quarterly review is a defensive exercise. The CTO or Senior Engineering Leader walks in with a spreadsheet of activities.

- "We closed 400 tickets."
- "We hired 10 developers."
- "We kept uptime at 99.9%."

When the CEO asks, *"Why is the mobile app update late?"*, the explanation dives into technical dependencies that sound like excuses to non-technical ears. The CFO looks at the "headcount" line item and wonders if the same work could be done with 10% fewer people.

The *Value Stream Executive* has an entirely different conversation.

They walk into the room not with a list of activities, but with a map of value. They don't defend headcount; they demonstrate impact.

- "We improved our Flow Efficiency from 20% to 35%, which means we are releasing capital into the market two weeks faster than last quarter."
- "Our investment in AI-augmented testing has reduced our defect escape rate, lowering our Cost of Poor Quality (COPQ) by $500k."
- "We are shifting capacity from the 'Legacy Maintenance' stream to the 'New Market Growth' stream to align with the company's Q3 strategic goals."

This is not a technology conversation. It is a business conversation about *portfolio allocation.* It shifts the question from *"How much does IT cost?"* to *"How fast can the company monetize its ideas?"*

This shift is critical because executives are beginning to ask different questions. They are asking, *"How can technology help us grow faster?"* rather than *"How many developers do you need?"* When you answer with Flow Metrics and Value Stream or Product managed P&L, you stop being a vendor to the business and start being a driver of the company.

Designing Teams to Reflect Systems

One of the most underestimated leadership responsibilities in enabling this convergence is team design.

How you organize your teams shapes how your systems evolve and how value flows through them. At Parchment, when we transitioned from a monolith to microservices, we didn't just refactor code—we refactored our organization.

We learned that hierarchical models often perform best in environments that require high coordination and tight regulatory control. But in a fast-moving, service-oriented architecture, self-managed teams thrive. They need autonomy, rapid iteration, and continuous learning.

Where leaders get into trouble is when they force one model into the wrong context. A self-managed team can't thrive if every decision must pass through layers of approval. Conversely, a hierarchical structure quickly breaks down in a microservices world where local decision-making matters more than conformity.

The most successful engineering organizations I've seen are those that intentionally design both architecture and teams as one integrated system. They understand that structure is strategy.[5]

Beyond Transformation: The Infinite Game

The title of this chapter is "Your Journey Starts Here," not "Your Journey Ends Here." That is intentional.

There is no "End State" for transformation. The moment you believe you have "finished" your Agile transformation or your move to the cloud, you have already begun to decay.

The market is a moving target. New competitors will emerge. New technologies, like the agentic AI models emerging today, will disrupt the economics of software delivery.

The goal, therefore, is to build an adaptable organism. When you view your organization as a Value Stream, you stop seeing friction as a people problem and start seeing it as a system problem. You stop asking, "Why is this developer slow?" and start asking, "What in our system is preventing this developer from flowing?"

YOUR MONDAY MORNING CALL TO ACTION

So, where do you begin? When you walk into the office, log onto *Slack*, or sign into *Teams* next Monday, what do you actually *do*?

1. Audit Your Funding Model

Look at your current budget. Are you funding temporary projects with start and end dates, or are you investing in products or value streams through stable, long-lived teams? If you are still funding projects, start the conversation with Finance about moving toward a product-based funding model (Chapter 9).

2. Map the Friction

Don't just look at Jira tickets. Walk the value stream. Identify where work waits. Is it waiting for approval? Waiting for a test environment? Waiting for a decision? The wait time is your most significant opportunity for improvement.

3. Prepare for the Talent Divergence

Assess your team's readiness for the AI age. Are you still hiring for rote coding skills, or are you hiring for "Vibe

Coding"? Start building guardrails that let your teams safely experiment with modern AI coding assistants and agentic workflows. Then measure whether they improve Flow and reduce risk.

4. Change the Conversation

The next time a stakeholder asks, "When will this be done?", answer with, "Here is the outcome we are driving toward, here is the leading indicator of value, and here is our current confidence based on what we know right now." Shift the focus from Output (features) to Outcome (business impact), and from false certainty to transparent learning.

When leaders deliberately anchor work to outcomes, leading indicators, and probabilistic confidence, consistency improves. Trust follows naturally.

The Shift That Changes Everything

Every transformation begins with awareness and ends with ownership.

The real work of change is not about adopting new frameworks or chasing the next methodology. It begins when leaders decide to think differently about technology, people, and purpose. That is where transformation moves from theory to reality.

I've seen both sides of the divide. I've lived through the age of control, when projects were managed through spreadsheets, approvals, and fear of mistakes, and I've witnessed the freedom that comes when teams are trusted, when data creates visibility, and when leaders let systems breathe.

The distance between those two worlds is measured in mindset, not tools. One treats technology as a cost to control. The other treats it as a force that can redefine the business itself.

Technology stops being a cost center the moment it proves its worth to the business. Every line of code, every platform investment, and every process improvement must tie back to one of three outcomes: profit, efficiency, or cost reduction.

And when technology is the product, as it is in modern SaaS companies, the stakes rise even higher. Technology does not just enable the business; it is the business.

Profit depends on building solutions that customers trust and value. Efficiency depends on shortening the path from idea to customer impact. Cost discipline depends on removing friction, waste, and rework from every step of delivery.

But no transformation sustains itself without leadership evolution. Systems follow the shape of their leaders. When authority moves to where information resides, when decisions occur closer to the work and the people performing it, organizations accelerate. The best leaders do not hoard control; they distribute it.

The lesson of Profitable Engineering is not just about financial return. It is about realization. It is about turning investment into impact and aligning innovation with intent.

- Profit funds purpose.
- Efficiency fuels innovation.
- Cost discipline creates freedom.

And at the center of all three stands the customer—because profit is sustained only when customers receive value worth paying for.

All work will always appear as a cost on the ledger. Salaries, systems, and software never disappear from the expense column. The leadership choice is whether technology remains defined as overhead or is managed as a partner in profit.

That is the essence of Profitable Engineering.

The future will be shaped by leaders willing to step forward and act.

Key Takeaways

- *Transformation Requires Business Alignment:* You can build a high-performing engineering organization, but if the business still views technology as a cost center, the impact will remain constrained. A Value Stream P&L helps connect delivery to business performance.
- *Culture Creates Capacity:* Deliberate investment in innovation builds the organizational muscle required to anticipate disruption and create advantage through it, including with AI.
- *Structure Is Strategy:* Team design should reflect architecture and flow. Autonomous, cross-functional teams are better suited to modern product delivery than functional silos.
- *The Convergence Is Key:* Flow, funding, realization, and AI are not separate conversations. Together, they form a modern executive operating model.
- *The Cycle Is Compressing:* What once took years now unfolds in months. Agility is no longer optional; it is a condition for staying relevant.
- Monday Morning Action: Stop funding projects. Start funding products and capacity. Stop measuring activity alone. Measure both how work flows and whether it delivers outcomes.

A Personal Reflection

I've spent more than two decades in this field, from writing my first lines of code to helping leaders see how technology can drive profit. Along the way, I've learned that the hardest distance to cover is not between idea and production. It is between potential and belief.

In every leader I have mentored and learned from and every transformation I've witnessed, it ultimately came down to one thing: someone believed there was a better way.

That belief carried me from the waterfall era of command-and-control into a world shaped by Agile, DevOps, and Value Stream Management. It led me to champion developer experience, measure flow, and embrace AI as a system enabler rather than a shortcut or gimmick.

And now, as I enter the next stage of my career, that same belief compels me to give back to the community, to the next generation of leaders, and to the industry that has given me so much.

This has been a story of growth, humility, and purpose. It is about learning to see technology not as machinery, but as a mirror that reflects who we are as leaders and what we choose to make possible.

So wherever you are in your journey, whether you are guiding a startup team or leading a global enterprise, remember this: reinvention is a mindset.

This book ends here, but the work of transformation does not. The next chapter is not mine to write. It is yours to live.

Part 3 Recap: The Business of Engineering

In this final section, we ascended from the engine room to the boardroom. We explored how to lead the convergence of human systems, intelligent tools, and business value:

What Changed in the Model

1. *Business–Technology Connection (Ch 9):* We stopped celebrating shipping and started closing the loop—verifying that our work changed customer behavior and business results.
2. *Leadership (Ch 10):* We evolved from task managers to market-makers—leaders who trade budget for speed and design the conditions where Flow and outcomes stay visible.
3. *AI (Ch 11):* We treated AI as a capacity multiplier—building literacy and guardrails so it accelerates Flow instead of amplifying chaos.
4. *The Journey (Ch 12):* We accepted there is no end state—transformation is a continuous discipline of reinvention.

Monday Morning Moves

- Walk into executive reviews with a map of value: outcomes, Flow signals, and what decisions those signals enable.
- Close the loop: require anticipated outcomes up front and revisit them after release to validate Realization—the evidence that delivered work produced the intended customer, operational, or business outcome.
- Lead investment, not activity: shift the conversation from "how much does IT cost?" to "how fast can we monetize our ideas?"
- Treat AI as a multiplier with guardrails: push literacy, establish constraints, and integrate AI where it compresses learning, not where it creates chaos.
- Keep the system from snapping back: defend the operating model deliberately when pressure rises.

Proof Signals

- You can produce evidence on demand: Flow signals (time, bottlenecks, reliability) alongside Realization signals (customer/financial impact).
- Executive conversations shift from headcount defense to portfolio allocation—because outcomes and tradeoffs are legible.
- Accounting remains a constraint, but not a stopping point; reporting stays compliant while leadership decisions are guided by outcomes, Flow, and a Value Stream P&L view.
- Cost-cutting decisions are debated as total economic impact (delay, risk, quality), not just expense line items.
- AI adoption shows up as system improvement (learning speed, safer execution), not tool-chasing.

The Work Ahead

The transformation is not "done." It is now a living operating model—and it must be defended, deliberately, against the pull of legacy reflexes.

Why it matters: Part 3 is where technology becomes legible in the language executives respect. When you address these pressures with discipline, the conversation shifts from defending cost to directing investment.

Every day, AI advances, and more organizations are experimenting with how to use it. It is difficult to predict what the next version of an organization will look like, or how AI will change the size and composition of teams, but the direction is clear: AI is already changing how we work, adapt, and deliver.

The organizations with the advantage will be the ones that embrace this shift and build true visibility into how work flows end to end across the enterprise. When you can see the flow, you can improve it. When you improve, you learn, adapt, and deliver outcomes faster.

In that world, technology cannot be treated as a back-office cost center. It must operate as a strategic partner, helping the business build the learning system, automation, and flow visibility required to turn AI from hype into a measurable advantage.

And it should not matter where you sit in the organization. When people understand their purpose, know the outcomes they own, and see how customers value the work they create, the work gains clarity. Collaboration gets easier. Delivery becomes more deliberate. Part 3 is ultimately about making that shift visible, measurable, and governable.

Appendix A: Bibliography

This bibliography consolidates sources cited within the chapters.

Chapter 1: The Cost Center Paradox

1. Drucker, Peter F. "Managing for Business Effectiveness." Harvard Business Review 41, no. 3 (May–June 1963).
2. Wiseman, Liz. Multipliers, Revised and Updated: How the Best Leaders Make Everyone Smarter. New York: HarperBusiness, 2017.
3. DevOps Research and Assessment (DORA). 2022 Accelerate State of DevOps Report. 2022. https://dora.dev/research/2022/dora-report/2022-dora-accelerate-state-of-devops-report.pdf.
4. IT Revolution. "Westrum's Organizational Model in Technology Organizations." May 5, 2021. https://itrevolution.com/articles/westrums-organizational-model-in-tech-orgs/.
5. Forsgren, Nicole, Jez Humble, and Gene Kim. Accelerate: The Science of Lean Software and DevOps. Portland, OR: IT Revolution Press, 2018.
6. Jaspan, Ciera, and Collin Green. "Measuring Productivity: All Models Are Wrong, But Some Are Useful." IEEE Software 42, no. 2 (2025): 13–17.
7. Pham, Kevin Phong, and Michael Neumann. "How to Measure Performance in Agile Software Development? A Mixed-Method Study." arXiv preprint arXiv:2407.06357, 2024.

8. Smart, Jon, Zsolt Berend, Myles Ogilvie, and Simon Rohrer. Sooner Safer Happier: Antipatterns and Patterns for Business Agility. Portland, OR: IT Revolution Press, 2020.
9. DHL Group. Human Resources and Social Report 2003. December 31, 2003. https://group.dhl.com/content/dam/deutschepostdhl/en/media-center/responsibility/8.4_EN_REP_HR%20and%20Social%20Report_2003.pdf.
10. FreightWaves. "Deutsche Post Acquires Airborne's U.S. Ground Operations for $1 Billion." March 25, 2003. https://www.freightwaves.com/news/deutsche-post-acquires-airbornes-u-s-ground-operations-for-1-billion.
11. Perri, Melissa. Escaping the Build Trap: How Effective Product Management Creates Real Value. Sebastopol, CA: O'Reilly Media, 2019.

Chapter 2: The Legacy Mirror

1. Toffler, Alvin. Future Shock. New York: Random House, 1970.
2. Gren, Lucas, and Niclas Pettersson. "Investigating Team Maturity in an Agile Automotive Reorganization." arXiv preprint arXiv:2409.11781, 2024. https://arxiv.org/abs/2409.11781.
3. Skelton, Matthew, and Manuel Pais. Team Topologies: Organizing Business and Technology Teams for Fast Flow. Portland, OR: IT Revolution Press, 2019.
4. Brooks, Fred. The Mythical Man-Month: Essays on Software Engineering. Anniversary ed. Reading, MA: Addison-Wesley, 1995.

5. Joseph, John, and Ronald Klingebiel. "Centralized Decision Making Helps Kill Bad Products." Harvard Business Review, October 18, 2016. https://hbr.org/2016/10/centralized-decision-making-helps-kill-bad-products.

Chapter 3: Practices That Drive Profitability

1. Emerson, Harrington. The Twelve Principles of Efficiency. New York: The Engineering Magazine Co., 1911.
2. Stone, Brad. The Everything Store: Jeff Bezos and the Age of Amazon. New York: Little, Brown and Company, 2013.
3. Forsgren, Nicole, Jez Humble, and Gene Kim. Accelerate: The Science of Lean Software and DevOps. Portland, OR: IT Revolution Press, 2018; Kersten, Mik. Project to Product: How to Survive and Thrive in the Age of Digital Disruption with the Flow Framework. Portland, OR: IT Revolution Press, 2018; Skelton, Matthew, and Manuel Pais. Team Topologies: Organizing Business and Technology Teams for Fast Flow. Portland, OR: IT Revolution Press, 2019; Smart, Jonathan, Zsolt Berend, Myles Ogilvie, and Simon Rohrer. Sooner Safer Happier: Antipatterns and Patterns for Business Agility. Portland, OR: IT Revolution Press, 2020.
4. Tuckman, Bruce W. "Developmental Sequence in Small Groups." Psychological Bulletin 63, no. 6 (1965): 384–99.
5. Kersten, Mik. Project to Product: How to Survive and Thrive in the Age of Digital Disruption with the Flow Framework. Portland, OR: IT Revolution Press, 2018.

6. Kersten, Mik. Project to Product: How to Survive and Thrive in the Age of Digital Disruption with the Flow Framework. Portland, OR: IT Revolution Press, 2018; Forsgren, Nicole, Jez Humble, and Gene Kim. Accelerate: The Science of Lean Software and DevOps. Portland, OR: IT Revolution Press, 2018.
7. Pereira, Steve, and Andrew Davis. Flow Engineering: From Value Stream Mapping to Effective Action. Portland, OR: IT Revolution Press, 2024.
8. Pereira, Steve, and Andrew Davis. Flow Engineering: From Value Stream Mapping to Effective Action. Portland, OR: IT Revolution Press, 2024.
9. Davis, Gwen. "Developer experience: What is it and why should you care?" GitHub Blog, June 8, 2023.
10. Forsgren, Nicole, Margaret-Anne Storey, Chandra Maddila, Tom Zimmermann, Brian Houck, and Jenna Butler. "The SPACE of Developer Productivity: There's More to It Than You Think." Queue 19, no. 1 (2021): 20–48.
11. Srivastava, Shivam, Kartik Trehan, Dilip Wagle, and Jane Wang. "Developer Velocity: How Software Excellence Fuels Business Performance." McKinsey & Company, April 20, 2020.

Chapter 4: Metrics That Matter

1. Goodhart, Charles A. E. "Problems of Monetary Management: The U.K. Experience." In Papers in Monetary Economics. Reserve Bank of Australia, 1975.
2. Snowden, David J., and Mary E. Boone. "A Leader's Framework for Decision Making." Harvard Business Review, November 2007.

3. Cunningham, Ward. "The WyCash Portfolio Management System." Addendum to the Proceedings on Object-Oriented Programming Systems, Languages, and Applications (OOPSLA), 1992.
4. Fowler, Martin. "Technical Debt." MartinFowler.com. October 1, 2003. Updated May 21, 2019. https://martinfowler.com/bliki/TechnicalDebt.html.
5. Reinertsen, Donald G. The Principles of Product Development Flow: Second Generation Lean Product Development. Redondo Beach, CA: Celeritas Publishing, 2009.
6. Doerr, John. Measure What Matters. New York: Portfolio, 2018.
7. Ellis, Sean, and Morgan Brown. Hacking Growth. New York: Crown Business, 2017.
8. Kersten, Mik, et al. Flow Metrics: A Business Leader's Guide to Measuring What Matters in Software Delivery. Planview, n.d.
9. Kersten, Mik. Project to Product: How to Survive and Thrive in the Age of Digital Disruption with the Flow Framework. Portland, OR: IT Revolution Press, 2018.
10. Forsgren, Nicole, Margaret-Anne Storey, Chandra Maddila, Tom Zimmermann, Brian Houck, and Jenna Butler. "The SPACE of Developer Productivity: There's More to It Than You Think." ACM Queue 19, no. 1 (2021).
11. McKinsey & Company. "Developer Velocity: How Software Excellence Fuels Business Performance." April 20, 2020.
12. Forsgren, Nicole, Jez Humble, and Gene Kim. Accelerate: The Science of Lean Software and DevOps. Portland, OR: IT Revolution Press, 2018.

13. Jaspan, Ciera, and Collin Green. "Measuring Productivity: All Models Are Wrong, But Some Are Useful." IEEE Software 42, no. 2 (2025): 13–17.
14. Edmondson, Amy C. The Fearless Organization. Hoboken, NJ: Wiley, 2018.
15. Google re:Work. "Understand Team Effectiveness." Re:Work. Accessed December 2025.
16. Atlassian. "Epics." Atlassian. Accessed December 2025.

Chapter 5: Redefining Product Operating Models

1. Drucker, Peter F. Management: Tasks, Responsibilities, Practices. New York: Harper & Row, 1973.
2. Kersten, Mik. Project to Product: How to Survive and Thrive in the Age of Digital Disruption with the Flow Framework. Portland, OR: IT Revolution Press, 2018; Skelton, Matthew, and Manuel Pais. Team Topologies: Organizing Business and Technology Teams for Fast Flow. Portland, OR: IT Revolution Press, 2019; McKinsey & Company. The Product-Based Operating Model: Unlocking Digital Transformation. New York: McKinsey & Company, 2023.
3. Cagan, Marty, and Felipe Castro. "Outcomes Are Hard." Silicon Valley Product Group, 2019.
4. Skelton, Matthew, and Manuel Pais. Team Topologies: Organizing Business and Technology Teams for Fast Flow. Portland, OR: IT Revolution Press, 2019.
5. Cagan, Marty. Transformed: Moving to the Product Operating Model. New York: Silicon Valley Product Group, 2024.

6. Evans, Eric. Domain-Driven Design: Tackling Complexity in the Heart of Software. Boston: Addison-Wesley, 2003.
7. Skelton, Matthew. "Lately, I've Been Thinking about How Easily Everything Gets Labelled a Product." LinkedIn post, December 10, 2025.
8. Cagan, Marty. Transformed: Moving to the Product Operating Model. New York: Silicon Valley Product Group, 2024.
9. Netflix Technology Blog. "Full Cycle Developers at Netflix—Operate What You Build." Netflix TechBlog, May 2018.
10. Netflix Technology Blog. "How We Build Code at Netflix." Netflix TechBlog, March 9, 2016.
11. Conway, Melvin E. "How Do Committees Invent?" Datamation 14, no. 4 (April 1968): 28–31.
12. Skelton, Matthew, and Manuel Pais. Team Topologies: Organizing Business and Technology Teams for Fast Flow. Portland, OR: IT Revolution Press, 2019.
13. McKinsey & Company. "The Big Product and Platform Shift: Five Actions to Get the Transformation Right." June 9, 2023.
14. McKinsey & Company. "The Bottom-Line Benefit of the Product Operating Model." December 19, 2023.
15. APQC. "Value Stream Mapping and End-to-End Process Improvement at Vanguard." February 6, 2020.
16. APQC. "Value Stream Mapping with the Vanguard Group." October 18, 2019.

Chapter 6: A New Way to Look at Leadership

1. Marquet, David. Turn the Ship Around!: A True Story of Turning Followers into Leaders. New York: Penguin Portfolio, 2013.
2. Conway, Melvin E. "How Do Committees Invent?" Datamation 14, no. 4 (April 1968): 28–31.
3. Skelton, Matthew, and Manuel Pais. Team Topologies: Organizing Business and Technology Teams for Fast Flow. Portland, OR: IT Revolution Press, 2019.
4. House Committee on Transportation and Infrastructure. The Design, Development, and Certification of the Boeing 737 MAX. Final committee report. Washington, DC: U.S. House of Representatives, September 2020.
5. Consumer Financial Protection Bureau. "Consumer Financial Protection Bureau Fines Wells Fargo $100 Million for Widespread Illegal Practice of Secretly Opening Unauthorized Accounts." September 8, 2016; Department of Justice. "Wells Fargo Agrees to Pay $3 Billion to Resolve Criminal and Civil Investigations into Sales Practices Involving the Opening of Millions of Accounts without Customer Authorization." February 21, 2020.
6. Lee, Jeremiah. "Spotify's Failed #SquadGoals." April 19, 2020.
7. Aghina, Wouter, Christopher Handscomb, Olli Salo, and Shail Thaker. "The Impact of Agility: How to Shape Your Organization to Compete." McKinsey & Company, May 25, 2021.
8. Scrum Inc. "Agile Unleashed at Scale: How John Deere's Global IT Group Implemented a Holistic Transformation Powered by Scrum@Scale, Scrum,

DevOps, and a Modernized Technology Stack." May 24, 2022.

9. Edmondson, Amy C. The Fearless Organization. Hoboken, NJ: Wiley, 2018; Google re:Work. "Project Aristotle." Re:Work, 2015; Forsgren, Nicole, Jez Humble, and Gene Kim. Accelerate: The Science of Lean Software and DevOps. Portland, OR: IT Revolution Press, 2018.

Chapter 7: Overcoming Barriers to Strategic Alignment

1. Meyer, Urban, with Wayne Coffey. Above the Line: Lessons in Leadership and Life from a Championship Season. New York: Penguin Press, 2015.
2. Kissler, Courtney. "Transforming to a Culture of Continuous Improvement." DevOps Enterprise Summit / LISA14 conference program, 2014.
3. Kim, Gene, Jez Humble, Patrick Debois, and John Willis. The DevOps Handbook: How to Create World-Class Agility, Reliability, and Security in Technology Organizations. Portland, OR: IT Revolution Press, 2016.
4. Forsgren, Nicole, Jez Humble, and Gene Kim. Accelerate: The Science of Lean Software and DevOps: Building and Scaling High Performing Technology Organizations. Portland, OR: IT Revolution Press, 2018.
5. Kersten, Mik. Project to Product: How to Survive and Thrive in the Age of Digital Disruption with the Flow Framework. Portland, OR: IT Revolution Press, 2018.
6. Allspaw, John. "Code as Craft." Etsy Code as Craft, 2012.
7. Wiseman, Liz. Multipliers: How the Best Leaders Make Everyone Smarter. New York: HarperBusiness, 2010.

8. Leopold, Klaus, and Siegfried Kaltenecker. Flight Levels: Leading Organizations with Business Agility. dpunkt, 2024.
9. McKinsey & Company. "ING's Agile Transformation." McKinsey Quarterly, 2017.
10. Scrum Inc. "Agile Unleashed at Scale: How John Deere's Global IT Group Implemented a Holistic Transformation Powered by Scrum@Scale, Scrum, DevOps, and a Modernized Technology Stack." May 24, 2022.
11. DORA Community of Practice. "How DORA Influenced John Deere's Agile Operating Model." 2023.
12. TBM Council. "How TBM & Product Costing Helped John Deere Accelerate Their Agile Transformation." Accessed March 23, 2026.

Chapter 8: Investing in People & Culture for High-Performing Teams

1. Sinek, Simon. Start with Why: How Great Leaders Inspire Everyone to Take Action. New York: Penguin, 2009.
2. Pink, Daniel H. Drive: The Surprising Truth About What Motivates Us. New York: Riverhead Books, 2009.
3. Instructure. "Instructure Signs Definitive Agreement to Acquire Parchment, the World's Largest Academic Credential Management Platform and Network." Press release, October 30, 2023.
4. Gallup. The Relationship Between Engagement at Work and Organizational Outcomes. Q12 Meta-Analysis, 11th ed. Gallup, May 2024. Updated July 2024.

5. Google re:Work. "Understand Team Effectiveness." Re:Work. Accessed March 24, 2026.
6. Westrum, Ron. "A Typology of Organisational Cultures." Quality and Safety in Health Care 13, suppl. 2 (2004): ii22–ii27.
7. Forsgren, Nicole, Jez Humble, and Gene Kim. Accelerate: The Science of Lean Software and DevOps. Portland, OR: IT Revolution Press, 2018.
8. Kim, Gene, Kevin Behr, and George Spafford. The Phoenix Project. Portland, OR: IT Revolution Press, 2013.
9. Smart, Jonathan, Zsolt Berend, Myles Ogilvie, and Simon Rohrer. Sooner Safer Happier: Antipatterns and Patterns for Business Agility. Portland, OR: IT Revolution Press, 2020.
10. Edmondson, Amy C. "Psychological Safety and Learning Behavior in Work Teams." Administrative Science Quarterly 44, no. 2 (1999): 350–83.
11. McKinsey & Company. "Developer Velocity: How Software Excellence Fuels Business Performance." 2020.
12. GitHub. The State of the Octoverse. 2023.
13. Noda, Ayako, Margaret-Anne Storey, and Laura Tacho. "DevEx: What Actually Drives Productivity." ACM Queue 21, no. 1 (2023).
14. Skelton, Matthew, and Manuel Pais. Team Topologies: Organizing Business and Technology Teams for Fast Flow. Portland, OR: IT Revolution Press, 2019.
15. Kotter, John P. Leading Change. Boston: Harvard Business Review Press, 1996.
16. Kerievsky, Joshua. "Modern Agile." Modern Agile. Accessed 2025.

Chapter 9: The Business-Technology Connection

1. Goldratt, Eliyahu M. The Goal: A Process of Ongoing Improvement. Great Barrington, MA: North River Press, 1984.
2. Kersten, Mik. Project to Product: How to Survive and Thrive in the Age of Digital Disruption with the Flow Framework. Portland, OR: IT Revolution Press, 2018.
3. DORA. "DORA's Software Delivery Performance Metrics." 2026.
4. Pereira, Steve, and Andrew Davis. Flow Engineering: From Value Stream Mapping to Effective Action. Portland, OR: IT Revolution Press, 2024.
5. Planview. The 2023 Project to Product State of the Industry Report. 2023; Broadcom. "The Top 7 Takeaways from the 2024 VSM Trends Survey." 2024.

Chapter 10: The Future of Engineering Leadership

1. Laozi. The Way of Life According to Laotzu: An American Version. Translated by Witter Bynner. New York: John Day, 1944.
2. Schmidt, Eric. "The AI Revolution Is Underhyped." TED2025, April 2025.
3. Clark, Phil. "Beyond Delivery: Realizing AI's Potential Across the Value Stream." Rethink Your Understanding, September 29, 2025.
4. Forsgren, Nicole, Jez Humble, and Gene Kim. Accelerate: The Science of Lean Software and DevOps: Building and Scaling High Performing Technology Organizations. Portland, OR: IT Revolution Press, 2018.

5. Atlassian. State of Developer Experience Report 2025. Atlassian, 2025.
6. Spotify. "What Is the Backstage Software Catalog?" Spotify for Backstage, 2024.
7. Skelton, Matthew, and Manuel Pais. Team Topologies: Organizing Business and Technology Teams for Fast Flow. Portland, OR: IT Revolution Press, 2019.
8. DORA. Accelerate State of DevOps Report 2024. Google Cloud, 2024.
9. LinearB. 2024 Engineering Benchmarks Report. LinearB, 2024.
10. Planview. The 2024 Project to Product State of the Industry Report. Planview, 2024.
11. Spring, Heather. "The Top 7 Takeaways from the 2024 VSM Trends Survey." Broadcom Software Academy, January 30, 2024.

Chapter 11: AI in Technology: The Orchestration Edge

1. Li, Fei-Fei. Quoted in "We Need Effective Governance to Shape AI for Good." United Nations University, May 29, 2024. https://unu.edu/article/we-need-effective-governance-shape-ai-good.
2. Tacho, Laura. "Only Cost Centers Brag about Cost Savings." LinkedIn post, October 2025. https://www.linkedin.com/posts/lauratacho_only-cost-centers-brag-about-cost-savings-activity-7378319585311260672-KTdj.
3. Tacho, Laura. "What the 2025 DORA Report Means for Your AI Strategy." DX Newsletter, October 2025.
4. Atlassian. State of Developer Experience Report. Atlassian, 2025.

5. FINOS. "Open Source Auto-Refactoring Meets AI Agent to Modernize Fintech Software at Scale." FINOS Blog, 2025.
6. Business Insider. "Morgan Stanley's DevGen.AI Tool Saved 280,000 Developer Hours." Business Insider Technology, 2025.

Chapter 12: Your Journey Starts Here

1. D'Souza, Francisco. Quoted in "Mint: Cognizant CEO Talks About His Work Ethic and Vision for the Future." Cognizant Newsroom, March 21, 2013.
2. Conway, Melvin E. "How Do Committees Invent?" Datamation 14, no. 4 (April 1968): 28–31.
3. Instructure. "Instructure Signs Definitive Agreement to Acquire Parchment, the World's Largest Academic Credential Management Platform and Network." Press release, October 30, 2023; PR Newswire. "Instructure Signs Definitive Agreement to Acquire Parchment..." October 30, 2023.
4. Skelton, Matthew, and Manuel Pais. Team Topologies: Organizing Business and Technology Teams for Fast Flow. Portland, OR: IT Revolution Press, 2019.

Appendix B: Profitable Engineering in a Vendor-Heavy World

Most senior leaders do not operate with a clean, fully in-house delivery model. They rely on some mix of contractors, systems integrators, offshore capacity, managed services, and strategic partners.

Sometimes that mix is intentional. Sometimes it emerges from cost pressure, hiring constraints, legacy platforms, acquisitions, or speed-to-market demands.

External capacity is not inherently the issue. The real risk appears when the operating model treats external work as outside the system, measures it differently, governs it differently, and integrates it late. Handoffs multiply, wait states grow, quality becomes negotiable, and Flow becomes fragile.

Profitable Engineering still applies in a vendor-heavy environment, but only when leadership refuses the most common illusion:

You can outsource capacity. You cannot outsource accountability.

The Principle: Outsource Capacity, Not Ownership

When technology is treated as a cost center, external work is often framed as a purchasing decision: "How do we buy output cheaper?"

That mindset breaks down under real delivery conditions because the business ultimately does not need output. It needs Realization—measurable change in customer behavior and business results. Ownership, therefore, must remain within the value stream.

- Internal leaders own the outcome.
- Internal teams own the architecture direction and the paved road.
- External teams can add capacity and execution, but strategy, quality, and accountability cannot be delegated away.

That is the difference between using partners to strengthen a value stream and using them in ways that create distance from ownership.

Designing Value Streams that Include Partners

If a value stream is the unit of Flow and Realization, partner work has to be designed into the value stream rather than bolted on at the end. The shift sounds subtle, but it changes the operating model:

- A single value stream map.
- A single backlog.
- A single definition of done.
- A single set of steering signals.

When partner teams are treated as separate systems, dependencies accumulate at the seams. Work waits longer, learning slows down, and accountability diffuses.

When partner teams are integrated into the same system, leaders can govern the work with the same discipline they apply internally.

Contracts Are Incentive Design

In a vendor-heavy world, the contract becomes a hidden operating model.

This is one of the places where transformations often lose coherence. Leaders try to run long-lived, product-oriented value streams while procurement structures still reward fixed scope, fixed deadlines, and "successful delivery" defined as handoff.

The behavior that follows is predictable: optimize for contract completion rather than customer outcomes.

A healthier approach is to treat contracting as incentive design—aligned to Flow and Realization rather than activity. Without getting lost in legal structures, the operating aim is straightforward:

- Protect continuity so learning compounds.
- Reduce handoffs so Flow tightens.
- Make quality non-negotiable so rework does not become the business model.
- Keep decision-making close to the work so the system does not drown in escalation.

No contract makes this perfect. But every contract either reinforces or undermines the operating model.

The Paved Road Applies to Partners Too

If you want external capacity without external chaos, partners need a safe path inside your system:

- Guardrails over gates
- Automation over approvals
- Default security and compliance over "we'll fix it later"
- Shared delivery standards over local improvisation

Partners should not have to invent their own delivery system inside yours. That is how variability turns into operational risk

A paved road reduces variability, speeds onboarding, and makes good delivery practices easier to repeat. It reduces uncertainty, accelerates onboarding, and makes outcomes more repeatable.

Measure the System, Not the Supplier

In cost-center thinking, vendor performance is often measured with activity metrics:

- tickets closed
- utilization
- deliverables completed
- hours burned

Those metrics are easy to report—and easy to exploit.

Profitable Engineering measures what survives contact with reality:

- *Flow:* How long does it take to get work through the system end-to-end, including vendor handoffs and wait states?
- *Quality:* What is the defect escape rate, rework load, and operational stability after release?
- *Realization:* Did the investment change customer behavior or business performance?
- *Health signals:* Is the system becoming easier to change or harder to change?

This is also why vendor management cannot sit in a separate governance lane. If external work is part of the value stream, it should be visible in the same steering dashboard. The goal is not simply to monitor a supplier. The goal is to steer the system.

Cost Pressure Is Where the Model Is Tested

When financial pressure hits, externalization becomes an easy lever. The expense is visible. The downstream cost is not.

This is the moment leadership has to act like a Value Stream Executive.

If a cost-cutting move slows Flow, increases rework, fractures coordination, or dilutes ownership, the business may reduce short-term expense while increasing long-term delay and risk. That is not savings. It's a shift from visible cost to invisible cost.

Profitable Engineering doesn't argue against partners. It argues against pretending the economics are simple.

Putting It into Practice

Use this checklist when external work is a meaningful part of your delivery model:

1. *Name the value stream owner.* Outcomes require a single accountable leader. External capacity can execute, but ownership stays inside.
2. *Buy teams, not bodies.* When you need external help, procure cross-functional capacity aligned to a value stream rather than isolated individual roles that create more seams.
3. *Unify the work system.* Use one backlog, one definition of done, one software development lifecycle (SDLC), and one release standard. Do not create a parallel delivery universe.

4. *Make handoffs visible.* Map where work waits across organizational boundaries. The biggest delays usually live in the seams.
5. *Use one steering dashboard.* Review Flow, quality, and Realization signals across the entire system, including partner work.

Case in Point: Buying teams, not bodies (2024)

In one recent example, we needed to augment capacity for what were, in effect, one-time initiatives. The reflex would have been straightforward staff augmentation: bring in individual engineers and push tasks across the fence.

Instead, we took a value stream approach.

What we actually needed was cross-functional, value-stream-aligned capacity with the right mix of roles to deliver end-to-end outcomes. We found a partner that could provide team-shaped capacity rather than individual staff augmentation.

Onboarding was bumpy at first because we had never taken this approach before, and our earlier experience had been almost entirely traditional staff augmentation. There was real learning friction in the beginning.

But once we integrated those teams into our system—our delivery practices, our software development lifecycle, code review expectations, and automation standards—the model started working. Even across time zones, the teams could follow the same practices, and we could govern the work within the value stream rather than through a separate vendor lane.

Both sides learned a great deal from the experiment. Most importantly, it reshaped our understanding of augmentation. We were not adding disconnected output. We were adding cohesive, outcome-capable capacity that could operate inside our paved road.

It is an approach I would use again.

The highest initial cost was onboarding teams to a shared system.

KEY TAKEAWAYS

- External capacity becomes risky when it is treated as outside the system.
- Accountability for outcomes must stay inside the value stream.
- Contracts shape behavior. When they reward handoffs, handoffs multiply.
- The paved road must include partners so variability does not become risk.
- Measure the system end to end. Metrics should change decisions, not decorate dashboards.

Recommended Reading

- Reinertsen, Donald G. *The Principles of Product Development Flow: Second Generation Lean Product Development*. Celeritas Publishing, 2009. Useful for understanding the economics of flow, queueing, and delay.
- Kersten, Mik. *Project to Product: How to Survive and Thrive in the Age of Digital Disruption with the Flow Framework*. IT Revolution, 2018. Helpful for framing value streams, flow, and product-oriented operating models.
- Skelton, Matthew, and Manuel Pais. *Team Topologies: Organizing Business and Technology Teams for Fast Flow*. IT Revolution, 2019. Strong reference for team design, interaction modes, and reducing coordination drag.
- Forsgren, Nicole, Jez Humble, and Gene Kim. *Accelerate: The Science of Lean Software and DevOps*. IT Revolution, 2018. Useful for connecting technical practices to delivery and organizational performance.
- Kim, Gene, Jez Humble, Patrick Debois, John Willis, and Nicole Forsgren. *The DevOps Handbook*. IT Revolution, 2016. Practical reference for delivery practices, automation, and operating discipline.

Appendix C: Value Stream Identification Template

An Optional Worksheet for Identifying and Discussing Value Streams

This appendix is intentionally practical. It reflects one way we approached identifying products and value streams using a combination of domain thinking, inputs and outputs, customer and business signals, and other patterns inside the system.

It is not presented as the only way to do this work, but as a useful example of how one organization made the exercise more concrete.

Readers looking for broader companion disciplines and suggested reading should see Appendix D, *Suggested Reading and Companion Practices for Modern Digital Product Delivery*. That appendix is designed as the "go deeper here" appendix rather than another template.

Value Stream: The sequence of activities required to deliver value from initial concept through customer delivery and operation.

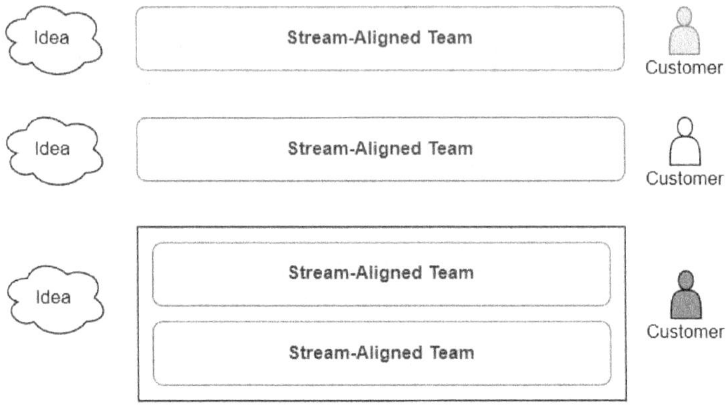

This appendix is a practical tool, not a formal methodology. It is designed to help leaders, architects, product teams, and engineering teams think more clearly about how value moves through a product or service.

The objective is not to produce a perfect taxonomy. It is to make the work visible enough to support better conversations about scope, ownership, dependencies, and improvement.

In many organizations, value stream identification becomes clearer when examined alongside architecture, product boundaries, and, where useful, Domain-Driven Design (DDD). These lenses can help teams reason about where value begins, where it is realized, and where dependencies slow delivery. They are helpful guides, not rigid requirements.

In some environments, a primary value stream may represent a broader product or product family, while sub-value streams may represent distinct capabilities within it. Those smaller streams may align with sub-domains, platforms, or enabling services. The right structure depends on the business model, architecture, and operating model.

Use this template as a starting point and adapt it to your context.

IDENTIFICATION OUTLINE
[Value Stream Name]

- Customer: [Who benefits the most directly from this value stream? E.g., end-users, internal teams, external clients.]
- Value delivered: [What outcomes or benefits does this value stream provide to its Customer? E.g., improved workflows, enhanced user experience, operational support.]
- Scope: [Define the boundaries and activities encompassed by this value stream. What steps or processes are included, and where does it begin and end?]
- Product: [Specify the product or suite of products associated with this value stream.]
- Dependencies:
    - Input Dependencies (Supplier): [List value streams or products this value stream depends on to function effectively.]
    - Output Dependencies (Buyers, Users): [List value streams or products that depend on this value stream.]
- Team: [Identify the team or teams responsible for executing this value stream.]

Example Value Streams

Example 1: Primary Value Stream — Signup-to-First Value (Customer Activation)

[Signup-to-First Value (Customer Activation)]

Customer: New customers (buyers/admins) and end users adopting a subscription-based SaaS product across any vertical.

Value delivered: A verified account, secure access, and a provisioned workspace that enables the customer to achieve their first meaningful outcome ("first value") with minimal friction.

Scope: Begins at initial sign-up (trial, self-serve purchase, or sales-assisted invite) and ends when the customer completes the first-value event (e.g., first record created, first workflow completed, first integration connected, first report generated). Includes registration, identity/login, tenant/workspace provisioning, baseline configuration, onboarding guidance, and instrumentation for activation.

Product: A multi-tenant B2B SaaS platform (web app + admin console + public API) with configurable workflows and reporting.

Dependencies

Input Dependencies (Supplier): Marketing/website or CRM invite, email/SMS delivery, identity provider (SSO/MFA), billing/commerce (if paywall), entitlements service, provisioning/infrastructure platform, analytics/telemetry.

Output Dependencies (Buyers, Users): Customer Success handoff, Support/Incident Response, Revenue Ops/Finance, Product analytics/experimentation, Security & compliance reporting.

Team: Growth & Activation team (PM, design, engineering) partnered with Identity Platform and Commerce/Billing platform teams.

Example 2: Sub-Value Stream — Registration & Account Creation

[Registration & Account Creation]

Customer: Prospective users and customer admins creating a new account/organization (trial, purchase, or invitation).

Value delivered: A created and verified user + organization/tenant identity, with the minimum profile data required to enter the product safely.

Scope: Begins when a user submits registration details and ends when the account is created and verified (email/domain verification, fraud/bot protection, initial profile). Excludes ongoing user management and billing.

Product: Registration services (sign-up UI/API, verification, account store).

Dependencies

Input Dependencies (Supplier): Email delivery, fraud/bot protection, customer/lead records (if sales-assisted), user/account database.

Output Dependencies (Buyers, Users): Identity & Login, Tenant Provisioning, Subscription & Entitlements, Support.

Team: Growth Engineering / Acquisition team.

Example 3: Sub-Value Stream — Identity & Login (Authentication, Authorization, Sessions)

[Identity & Login]

Customer: End users and customer IT administrators who require secure access (SSO/MFA), least-privilege roles, and auditability.

Value delivered: Reliable authentication and authorization that grants the right user the right access—quickly, securely, and with traceable audit evidence.

Scope: Begins at a login attempt (password, SSO, magic link) and ends when a valid session/token is established with correct entitlements and an audit event recorded. Includes MFA/SSO flows, session policy, RBAC enforcement, and access logging.

Product: Identity platform (SSO, MFA, RBAC, token/session services, audit logs).

Dependencies

Input Dependencies (Supplier): User/account records, customer IdP metadata, secrets/cert management, policy configuration, observability/logging.

Output Dependencies (Buyers, Users): All product value streams that require access; Support (recovery/lockouts); Compliance reporting; Admin security controls.

Team: Identity Platform team (platform engineering) partnered with Security and SRE.

Example 4: Sub-Value Stream — Subscription Order Processing (Quote-to-Cash / Order-to-Cash)

[Subscription Order Processing]

Customer: Buyers (admins/procurement) purchasing, upgrading, or renewing a subscription—self-serve or sales-assisted.

Value delivered: An executed order/subscription with accurate pricing and an entitlement record that controls access to product capabilities.

Scope: Begins when a plan/offer is selected (or a quote is accepted) and ends when the subscription/order is created and entitlements are issued. Includes plan catalog, pricing rules, discounts, tax/VAT inputs, and entitlement creation. Excludes invoice collection and revenue recognition.

Product: Commerce services (plan catalog, check-out/ordering, subscription management, entitlements).

Dependencies

Input Dependencies (Supplier): Plan catalog/pricing, CRM/CPQ (if sales-assisted), tax calculation, identity (buyer authentication), payments/billing platform.

Output Dependencies (Buyers, Users): Billing & Payments, Tenant Provisioning, Customer Success, Revenue reporting, Support.

Team: Commerce/Monetization team partnered with Finance Systems.

Appendix D: Suggested Reading and Companion Practices for Modern Digital Product Delivery

This book was never intended to be a comprehensive implementation manual for every practice supporting modern digital product delivery. That was a deliberate choice.

The goal of *Profitable Engineering* is to help leaders see the larger system more clearly: how customer needs, strategic positioning, domain clarity, architecture, team design, flow, and outcomes connect.

For readers who want to go deeper, however, it helps to point toward a set of companion disciplines that strengthen that larger picture without turning this book into a separate field guide for each one.

I recommend these practices as a layered approach to modern product delivery, not as a collection of disconnected tools.

One discipline helps leaders understand what users actually need. Another helps them see the strategic landscape around those needs. Another clarifies the business domain and its boundaries. Another makes architecture visible enough to discuss. Another helps shape teams for fast flow. Another helps leaders measure whether the whole system is actually improving. The value is not in adopting every idea mechanically.

The value is in understanding how the practices reinforce one another when used with intention.

User Needs Mapping

A useful place to begin is with User Needs Mapping, especially through Rich Allen's *User Needs Mapping: Aligning Teams Around What Matters.*[1] This work is a strong fit for leaders trying to reconnect delivery systems to purpose.

Allen describes User Needs Mapping as a way to align teams around real user outcomes, reduce dependencies, improve organizational clarity, and make the invisible visible. His framing is especially helpful because it does not stop at discovery language. It extends into capabilities, team interactions, cognitive load, and flow, which makes it more relevant to operating-model thinking than many narrower user-research methods.[1]

In a broader strategic approach, User Needs Mapping helps prevent one of the most common forms of drift in technology organizations: the inside-out tendency to organize around systems, silos, and inherited structures rather than around what people are actually trying to achieve.

That makes it a strong starting point for this appendix. Before leaders debate architecture, metrics, or org charts, they need a clearer view of what matters to users and where work gets distorted on its way from intention to delivery.

Wardley Mapping

Once user needs are clearer, the next question becomes strategic: where should the organization differentiate, where should it standardize, and what is shaping the landscape around those choices?

Wardley Mapping is valuable here because it introduces a visual, situationally aware way to think about strategy. Simon Wardley's material describes it as a visual approach to strategy that helps people communicate how things work, where the problems are, and what they intend to do about them when decisions matter.[2]

In the context of a total strategic approach, Wardley Mapping adds something different from user-needs work. It helps leaders reason about context, movement, inertia, and competitive posture. It sharpens decisions about what is truly differentiating, what has become more utility-like, and where energy is being spent in the wrong places. That makes it especially useful for senior leaders who need a bridge between customer-centric thinking and investment choices.

Domain-Driven Design

If Wardley Mapping sharpens strategic position, *Domain-Driven Design* helps sharpen meaning. Eric Evans' work remains one of the most useful references for teams dealing with real business complexity.[3]

The official Domain Language materials describe DDD as a broad framework for making design decisions and a vocabulary for discussing domain design. At the same time, the DDD Reference serves as a summary of the patterns and definitions that underpin the discipline.[3]

This matters in a total strategic approach because organizations often move too quickly from priorities to implementation without ever becoming precise about the domain itself.

When business language is vague, boundaries become vague. When boundaries become vague, ownership, architecture, and delivery all suffer. DDD helps create a clearer connection between product intent and software design by making the domain discussable in a disciplined way. That is one reason it sits naturally between strategy and architecture in this sequence.

Architecture Visibility and the C4 Model

One of the recurring problems in software organizations is that architecture is often treated as important without being made visible in a consistent, useful way. The C4 model addresses that problem directly.[4]

Simon Brown describes it as an easy-to-learn, developer-friendly approach to software architecture diagramming built around hierarchical abstractions such as software systems, containers, components, and code. The model is practical precisely because it avoids both hand-waving and unnecessary complexity.[4]

In a broader operating model, the C4 model helps architecture become part of the strategic conversation instead of a private language spoken only by architects. Once teams have a clearer grasp of user needs, strategic position, and domain boundaries, they still need a way to visualize the system at the right level of detail.

That visibility improves conversations about dependencies, modernization, boundaries, and technical risk. It also helps architecture support flow rather than obscure it.

Team Topologies

If architecture makes the system more visible, *Team Topologies* helps make the organization more intentional.[5]

The official Team Topologies material defines it as an approach to designing team-of-teams organizations for the fast flow of value. It centers on four fundamental team types and a set of interaction modes designed to improve autonomy, collaboration, delivery focus, and product alignment.[5]

This is where many operating-model conversations become real. Leaders can understand users, talk strategy, clarify domains, and document architecture, yet still fail if the organization remains trapped in poorly designed handoffs, overloaded teams, and ambiguous ownership.

Team Topologies helps address that by treating team design as a first-order concern rather than an afterthought. It gives leaders a more deliberate way to align team boundaries, interaction patterns, and platform thinking to improve flow.

Value Stream Management and Flow Engineering

Value Stream Management belongs near the end of this progression, not because it is less important, but because it becomes more meaningful when the earlier layers are clearer.

The Flow Framework Community describes the Flow Framework as a blueprint for implementing Value Stream Management, connecting IT and the business, and helping organizations develop expertise in Value Stream Management and Flow Metrics.[6]

The *Project to Product* material frames that shift around moving from project thinking toward a value stream network and product-centric management.[6]

I would also include *Flow Engineering: From Value Stream Mapping to Effective Action* by Steve Pereira and Andrew Davis alongside Mik Kersten's work.[7]

The official Flow Engineering site describes the book as a guide to an open set of adaptive practices for eliminating waste and improving performance across organizations. At the same time, the publisher's summary presents it as a practical guide to using value stream mapping techniques to align teams, unlock innovation, and optimize performance.[7]

Taken together, these books complement one another well: Kersten provides the management and measurement lens for Value Stream Management, while Pereira and Davis provide a more action-oriented path from mapping to improvement. That pairing makes the section stronger.[6,7]

In a total strategic approach, this is where the loop closes. The earlier practices help leaders understand what matters, make better strategic choices, clarify business boundaries, visualize architecture, and shape teams around flow. Value Stream Management and Flow Engineering help leaders see whether the whole system is actually improving and where it is not. They create visibility into movement, bottlenecks, dependencies, delays, and improvement opportunities across the system rather than inside a single function. That is why they belong together in the appendix.

Putting the Practices Together

What makes these practices valuable is not that each one is powerful on its own, though many are. It is that together they help leaders build a more coherent operating model.

User Needs Mapping anchors the work in what matters to users. Wardley Mapping adds strategic awareness. Domain-Driven Design brings clearer language and boundaries. The C4 model makes architecture visible enough to reason about. Team Topologies shape the organization for the fast flow of value. Value Stream Management and Flow Engineering help leaders see whether the whole system is learning, improving, and delivering better results over time.

That is why I would frame this appendix as suggested reading and companion practices, not as a checklist or a mandated stack. A single framework, reorganization, or dashboard does not solve modern digital product delivery. It improves when leaders can connect customer needs, strategic posture, domain clarity, architecture, team design, and flow visibility into a more integrated system of work.

References:

1. Allen, Rich. *User Needs Mapping: Aligning Teams Around What Matters.* 2025.
   Accessed November, 2025. https://userneeds-mapping.com/book-orders.

   Allen, Rich. *"User Needs Mapping."*
   Accessed November, 2025. https://www.userneeds-mapping.com/.

2. Wardley, Simon. *Learn Wardley Mapping.*
   Accessed December, 2025. https://learnwardleymapping.com/.

3. Evans, Eric. *Domain-Driven Design: Tackling Complexity in the Heart of Software.* Boston: Addison-Wesley Professional, 2004.

   Evans, Eric. *DDD Reference: Definitions and Pattern Summaries.* Domain Language, Inc., 2015.
   Accessed October, 2025. https://www.domainlanguage.com/ddd/reference/.

4. Brown, Simon. *The C4 Model for Visualising Software Architecture.*

   Accessed October, 2025. https://c4model.com/.

5. Skelton, Matthew, and Manuel Pais. *Team Topologies: Organizing Business and Technology Teams for Fast Flow.* Portland, OR: IT Revolution Press, 2019.

"Team Topologies."

Accessed November, 2025. https://teamtopologies.com/.

6. Kersten, Mik. *Project to Product: How to Survive and Thrive in the Age of Digital Disruption with the Flow Framework.* Portland, OR: IT Revolution, 2018.

"Flow Framework Community."

Accessed December, 2025. https://flowframework.org/.

7. Pereira, Steve, and Andrew Davis. *Flow Engineering: From Value Stream Mapping to Effective Action.* Portland, OR: IT Revolution Press, 2024.

"Flow Engineering."

Accessed October, 2025. https://flowengineering.org/.

www.ingramcontent.com/pod-product-compliance
Lightning Source LLC
LaVergne TN
LVHW052220200926
848481LV00001B/169
9798996116904